INDUSTRIAL EMERGENCY PREPAREDNESS

INDUSTRIAL EMERGENCY PREPAREDNESS

Robert B. Kelly

 Van Nostrand Reinhold
New York

Copyright © 1989 by Van Nostrand Reinhold

Library of Congress Catalog Card Number 89-5600
ISBN 0-442-20483-3

Printed in the United States of America

Van Nostrand Reinhold
115 Fifth Avenue
New York, NY 10003

Van Nostrand Reinhold International Company Limited
11 New Fetter Lane
London EC4P 4EE, England

Van Nostrand Reinhold
480 La Trobe Street
Melbourne, Victoria 3000, Australia

Nelson Canada
1120 Birchmount Road
Scarborough, Ontario   M1K 5G4, Canada

16 15 14 13 12 11 10 9 8 7 6 5 4 3 2 1

Library of Congress Cataloging-in-Publication Data

Kelly, Robert B., 1958–
   Industrial emergency preparedness / by Robert B. Kelly.
      p.   cm.
   Includes index.
      ISBN 0-422-20483-3
   1. Industrial safety. 2. Disaster relief. I. Title.
T55.K45   1989
670'.28'9—dc19   89-5600
                                                          CIP

DEDICATION

This book is dedicated to my wife, Debbie, whose patience, support, and advice in writing this book was invaluable; to my daughter, Michelle, for giving me motivation; and to my parents for giving me the opportunity and education to pursue my goals.

Contents

Preface

In 1986, I was a management consultant providing emergency management advice to a number of industrial clients. On behalf of one client, the American Society of Safety Engineers, I conducted several seminars entitled "Disaster and Emergency Preparedness." One of the students, a safety engineer with a major chemical company, suggested that there were not many practical books concerning industrial emergency preparedness. There were many written for government agencies, often from a highly theoretical or sociological viewpoint, but none based on practical experience that met industry's needs.

Shortly thereafter I decided to write this book. My goal was to provide useful information to industrial planners tasked with developing emergency preparedness plans and programs. In conveying emergency preparedness information, I drew on more than ten years of emergency management experience at the federal, state, and local levels of government, as well as on several years of working for a variety of U.S. and foreign industrial clients. I believe the people that will most benefit by this book are safety engineers, environmental engineers, plant managers, risk managers, loss prevention engineers, human resource managers, and industrial hygienists.

Usually when I read a book relative to my profession, I find that I read it in 15-to-20 minute spurts, usually over a cup of coffee in the morning before the hectic day begins. Figuring that you might have similar reading habits, I divided this book into many short chapters so that you could read a complete chapter during the breaks in the day that you might have. While the book can be read quickly, I believe it will be a good reference book to keep handy. It contains many useful charts, diagrams, and checklists that will aid you in the development of your facility's emergency management program.

Industrial emergency planning is becoming an increasingly important part of a company's risk management and loss prevention program. I hope this book provides you with the knowledge you need to develop effective emergency preparedness programs for your facility.

ROBERT B. KELLY

Acronyms

CAER	Community Awareness and Emergency Response
CEC	Community Emergency Coordinator
CEO	Chief Executive Officer
CHEMTREC	Chemical Transportation Emergency Center, operated by the Chemical Manufacturers Association
ECC	Emergency Control Center (a synonym for an EOC)
EHA	Extremely Hazardous Substance
EOC	Emergency Operations Center
EPA	Environmental Protection Agency
FEC	Facility Emergency Coordinator
FEMA	Federal Emergency Management Agency
LEPC	Local Emergency Planning Committee
MSDS	Material Safety Data Sheets
MSEL	Master Sequences of Events List (see Chapters regarding Drills and Exercises)
NFPA	National Fire Protection Association
NIOSH	National Institute of Occupational Safety and Health
NRC	Nuclear Regulatory Agency (Some people also refer to the National Response Center, a government center for reporting chemical emergencies, as the NRC, but this book's use of NRC generally refers to the Nuclear Regulatory Agency)
NRT	National Response Team
OSHA	Occupational Safety and Health Administration
RCRA	Reseource Conservation and Recovery Act

RQ   Reportable Quantity
SARA   Superfund Amendments and Reauthorization Act
SERC   State Emergency Response Commission
TPQ   Threshold Planning Quantity

INDUSTRIAL
EMERGENCY
PREPAREDNESS

1

The Need for Emergency Preparedness

WHAT IS EMERGENCY PREPAREDNESS?

Emergency (or disaster) preparedness encompasses all activities that are necessary to prepare people and organizations to respond to emergencies and disasters. These activities seek to facilitate the response to save lives and minimize damage to property in the event of emergency. Examples of these activities include: developing emergency plans and procedures, assembling equipment and other resources necessary to combat emergencies, training facility personnel, conducting drills and exercises, and developing and implementing public education programs.

Although the terms "emergency" and "disaster" are often used interchangeably, it is important to note the difference between the two. "Emergency" usually refers to "an unforeseen combination of circumstances or the resulting state that calls for immediate action." "Disaster" is defined as "a sudden calamitous event bringing great damage, loss, or destruction." Therefore, not all emergencies are disasters. **The degree and effectiveness of preparedness often spell the difference between emergency and disaster.**

Preparedness is just one phase of a comprehensive emergency management program. The other phases are prevention, response, and recovery. Prevention, the initial phase, is the practice of activities designed to prevent accidents and emergencies from occurring. Response follows preparedness and involves lifesaving and protection activities that are implemented during an emergency (e.g., spill control, firefighting, evacuation, etc.). The final phase, recovery, embodies all the activities necessary to bring the organization back to normal or routine operations.

These four phases are cyclical in nature, as shown in figure 1–1. Each phase evolves into the next. Of course there is nothing to suggest that one phase does not last much longer than another; after all, the point of prevention phase is to do all that can be done to prevent emergencies from happening. The prevention and preparedness phases can last years or decades before any actual emergency occurs. If an emergency were to occur, however, the response phase would lead to the recovery phase, which would be followed by an entirely new emergency management cycle, beginning with prevention.

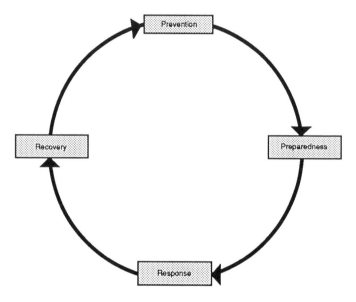

Figure 1–1: Cyclical nature of emergency management

The preparedness phase is itself composed of many elements. Some people mistakenly believe that an emergency plan is the sum and substance of a preparedness program. As important as it is, the emergency plan is but one element of an effective industrial emergency preparedness program, as illustrated in figure 1–2.

WHY IS AN EMERGENCY PREPAREDNESS PROGRAM NECESSARY?

Why should we be prepared to handle emergencies? One reason is obvious: Emergencies *can* and *do* occur, and *when* they occur, our natural instinct is to protect ourselves, and others, and our property. The only effective and logical way to do this is to prepare in advance

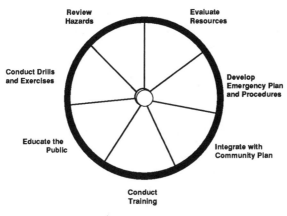

Review
Hazards

Evaluate
Resources

Conduct Drills
and Exercises

Develop
Emergency Plan
and Procedures

Educate the
Public

Integrate with
Community Plan

Conduct
Training

*The emergency plan is only one
part of an emergency preparedness program*

Figure 1–2: Elements of an emergency preparedness program

for such events. But beyond the obvious are other legitimate, intelligent reasons for developing an emergency preparedness program within any organization. Businesses today must be concerned with insurance stipulations, media attention, public pressure, employee health and safety, liability, and government regulations. Emergency preparedness is necessitated by all these external factors, as illustrated in figure 1–3.

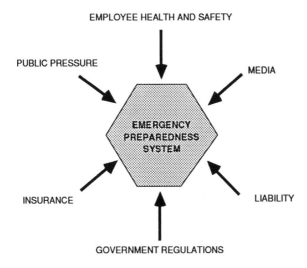

EMPLOYEE HEALTH AND SAFETY

PUBLIC PRESSURE

MEDIA

EMERGENCY
PREPAREDNESS
SYSTEM

INSURANCE

LIABILITY

GOVERNMENT REGULATIONS

Figure 1–3: External factors necessitating an effective emergency management program

All types of insurance—especially liability—can be very costly and sometimes difficult to obtain in today's business world. Because of this, many businesses are forced to either self-insure or improve their loss prevention programs—or both. Since the Bhopal incident, emergency preparedness programs have been looked at much more closely by potential insurers than they have been in the past. The media and public have also scrutinized these emergency preparedness programs, and as a result, if a business expects to operate hazardous processes within community environs, it had better be well prepared for emergency situations. And the ever-present roster of government safety regulations is another good incentive for responsible businesses to enhance their emergency preparedness programs.

The aforementioned considerations are born largely of negative inducements, but effective emergency preparedness programs are made desirable by positive motives and tangible benefits as well. For one thing, insurance coverage may be more readily available and/or more affordable to an organization with a strong emergency preparedness program. Also, studies have shown that organizations that utilize effective emergency response systems during emergencies suffer much lower losses than do comparable organizations with poor emergency response systems. In fact, loss figures for organizations with strong emergency response systems are sometimes as little as 6% of the loss figures for organizations with poor emergency systems. (See figure 1–4.)

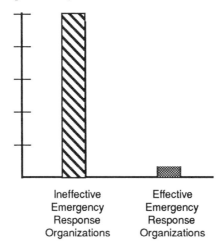

Ineffective   Effective
Emergency   Emergency
Response   Response
Organizations   Organizations

Studies have shown that companies with effective emergency response organizations suffer losses only 6% of those companies with ineffective response organizations

Figure 1–4: Losses as a function of effective emergency response organization

Within the past several years, there have been a number of lawsuits by members of the public against companies that have suffered disasters. A common theme of these lawsuits is poor implementation of emergency procedures. Even though today's jury awards can be staggering, the cost of legal defense alone can run into the millions of dollars. By developing an effective emergency preparedness program that includes adequate training to ensure proper implementation, many such lawsuits can be eliminated.

In summary, effective preparedness programs can more than pay for themselves in good public relations, potentially lower insurance and legal costs, and less regulatory pressure.

EMERGENCY PREPAREDNESS REGULATIONS

There are a number of federal and state laws that mandate the development of emergency plans for businesses; however, the applicability of those requirements often depends on the nature of the business operations. For example, a plant in which hazardous substances exist would require emergency plans specifically appropriate for the use, generation, and storage of those substances.

Federal Regulations

While statutory requirements for emergency planning may vary among industries, most businesses with more than 10 employees are required by the Occupational Safety and Health Administration (OSHA) to develop and maintain employee emergency and fire prevention plans. OSHA regulation 29 CFR 1910.38 outlines the requirements for the plan: The plan must cover "those designated actions employers and employees must take to ensure employees safety from fire and other emergencies." The plan must also contain emergency escape procedures, emergency shutdown procedures, personnel accountability, rescue and medical duties, and other emergency functions.

An example of industry-specific emergency planning requirements concerns businesses that deal with hazardous wastes. Under the Resource Conservation and Recovery Act (RCRA), companies that own or operate hazardous waste facilities are required to prepare contingency plans to mitigate health and environmental damage from fires; explosions; or any sudden or non-sudden release of hazardous waste to air, soil, or surface. These plans must be site-specific and are required prior to the issuance of a hazardous waste permit (also known as the RCRA Part B permit). Other industry-specific emergency planning requirements target owners of onshore and offshore oil facilities that might discharge harmful quantities of oil into navigable waters of the United States. Under Environmental Protection Agency (EPA) regulations, such companies are required to develop and maintain a "spill prevention control and countermeasure plan."

Regulation	Industries Affected	Compliance Difficulty
OSHA (29 CFR 1910.38)	Most	Low
RCRA (40CFR256.30)	Some	Moderate
SPCC (40CFR112.7)	Some	Moderate
CERCLA	Many	Moderate
SARA-Title III	Many	Low
NUREG 0654	Nuclear	High

Figure 1–5: Examples of federal regulations requiring emergency preparedness in industrial facilities

Figure 1–5 gives some examples of federal regulations that may or may not be applicable to your business situation. If you are unsure of which regulations might affect your facility, contact a lawyer or other knowledgeable people. Complying with the provisions of these laws is not difficult and is to your company's advantage to do so.

Another example of emergency preparedness regulation that affects your facility is Title III of the Superfund Amendments and Reauthorization Act (SARA). This law mandates a minimum level of community emergency planning for hazardous materials accidents and requires input from industry. A summary of the emergency planning requirements of this act follows.

EMERGENCY PLANNING: SECTIONS 301–303
The objectives of Title III are to improve local chemical emergency response capabilities (primarily through improved planning and notification) and to provide citizens with access to information about chemicals in their localities.

Title III addresses planning by (1) defining the extremely hazardous substances that trigger the planning process, (2) requiring the establishment of a state and local planning structure and process (including specifics on committee membership), (3) requiring facilities to make information available to local planners, and (4) specifying the minimum contents of a local emergency plan.

Title III requires the EPA to publish a list of extremely hazardous substances (EHSs) and threshold planning quantities (TPQs)

for each EHS. The EPA fulfilled this requirement in rules published on November 17, 1986 (51 FR 41570); April 22, 1987 (52 FR 13378-13410); December 17, 1987 (52 FR 48072-73); and February 25, 1988 (53 FR 5574-5). (See appendix B of the National Response Team's *Technical Guidance for Hazards Analysis* for an explanation of the criteria used in identifying those chemicals now referred to as EHSs).

PLANNING STRUCTURE AND PROCESS: SECTIONS 301–303

State governors must appoint State Emergency Response Commissions (SERCs) by April 17, 1987, and appoint members to the Local Emergency Planning Committees (LEPCs) by August 17, 1987. The SERCs are to coordinate and supervise the work of the LEPCs and review all emergency plans to ensure that all the local plans for any one state are coordinated.

Facilities must notify the SERCs by May 17, 1987, if they have any listed EHSs that exceed the designated TPQ. (The TPQ is a specific quantity assigned to each of the 366 extremely hazardous substances.) If a facility has present an EHS in an amount greater than the TPQ, the facility must identify itself to the SERC. The SERC will notify the LEPC so that the facility can be included in the emergency planning process.

Facilities must provide the following information to the LEPC:

- the name of the facility representative (by September 17, 1987) to serve as emergency coordinator and to assist the LEPC in the planning process
- information requested by the LEPC that is necessary for developing and implementing the emergency plan (see Section 303(d)(3) of SARA)
- any changes at the facility that could affect emergency planning

LEPCs must prepare comprehensive emergency plans by October 17, 1988, and update them annually. The planning infrastructure for receiving information and formulating plans is graphically displayed in NRT-1, Exhibit 1-1. A summary of key Title III dates affecting planning can be found in Exhibit 1-2 of NRT-1.

State Laws

Several states are also beginning to require companies to develop emergency plans. Many states are passing right-to-know laws that often contain provisions for businesses to develop emergency plans in conjunction with local officials. Some states separate right-to-know laws from the planning laws. Notable among these state laws are the Illinois Chemical Safety Act and the New Jersey Toxic Catastrophe Prevention Act. The Illinois law requires many industrial companies that deal with hazardous materials to develop emergency response plans and to submit them to their local disaster services agencies. Training of employees in emergency response procedures is a major element of this comprehensive piece of legislation. The New Jersey law is aimed primarily at the prevention of catastrophic accidents rather than at merely preparing for them. Thus, its requirements revolve more around activities such as risk assessment and process control, but it still requires facilities to have comprehensive risk management programs that include detailed emergency response procedures and plans.

By checking with your company's attorney, state or local emergency preparedness office, health department, or environmental protection division, you should be able to determine your legal requirements for developing emergency plans.

PREPARING FOR THE INEVITABLE

While everyone knows that emergencies and disasters can strike anywhere and anytime, there is no law of nature that mandates that a disaster must strike your facility. Good emergency management techniques can help you avoid the so-called "inevitable." But you must be prepared to open your mind to the multitude of hazards that are possible. There is no such thing as "it can't happen here." It can and probably will unless you take positive action to preclude it. Chapters 2 and 3 discuss a variety of natural and technological hazards that can affect your facility, but by no means should such a list be considered complete. There are countless factors that can complicate hazards, and strange occurrences that one could not anticipate without an open mind. Consider the following story about a highly unlikely, yet tragic, flood.

> *Boston, Massachusetts. January 15, 1919:* Disaster occurred on one of those rare "June-in-January" days experienced in New England when tropical breezes blow up the East coast. Residents of Boston's North End, many of them recently-arrived Italian immigrants; plus scores of workers from nearby factories, warehouses, and docks, sat outside enjoying the midday sun. Towering over this neighborhood of tenements and historic sites was a 90-foot-high metal tank, 282 feet in circumference. Standing at the corner of Foster and Commercial streets, the giant vat was used for the storage of raw molasses by the Purity Distilling Company.

With a low rumbling noise, followed by a series of sharp explosions, the tank burst open and a black flood of molasses poured into the streets. Faster than anyone could run, a wave of sticky syrup, initially 20–30 feet high, flowed through the narrow streets, burying workmen, pedestrians, and lunchtime idlers. An estimated 2 million gallons of molasses, weighing some 27 million pounds, had been released from the burst tank. The flood knocked several buildings from their foundations and drowned or suffocated 21 people where they stood. Sections of the ruptured tank sliced through building walls and sheared off columns supporting an elevated train line.

Would-be rescuers and sightseers found themselves wallowing in a knee-deep sludge. Survivors had to be cut from their sugar-encrusted clothing. Horses became hopelessly trapped in the syrup and had to be shot by police. By the end of the day, most of the city seemed sugar-coated, as visitors to the scene tracked the molasses residue wherever they went. The odor of molasses lingered over the city for a week, and the harbor remained brown-tinged for almost five months.

Several theories were proposed for the burst of the molasses tank. Some authorities felt the warm weather had caused the molasses to expand and rupture the tank seams. Spokesmen for the Purity Distilling Company claimed at first that vibrations from a passing train had caused the collapse. Later they tried to suggest that anarchists (at that time believed to be numerous among the North End's large Italian population) had blown it up. The official investigation instead found the company guilty of shoddy workmanship in the original construction of the tank and ordered Purity Distilling to pay over one million dollars in damages.

The point of telling this sad, strange story is to exemplify the boundless possibilities of unexpected occurrences. Many tragic and costly things can happen at your plant; an open mind and flexible planning can help to prevent them.

2

Natural Hazards

While there are a multitude of disaster-causing hazards in the world, it is possible to categorize these hazards. Hazards can be classified as either **natural** or **technological**. Technological hazards, sometimes referred to as man-made hazards, are usually events caused by man's negligence in handling his technology. This negligence can be the result of human error, tiredness, poor maintenance of equipment, or lack of adequate training. Natural hazards are events that are caused by the forces of nature such as drought, flood, hurricane, tornado, earthquake, and winter storm.

DROUGHT

Nature and Cause

Drought is a condition of dryness, where soil, water bodies, and vegetation are depleted of moisture that is not replenished in adequate quantities. Droughts are often gradual developments and may last for years.

Droughts result from several meteorological factors. Lack of rainfall is central, but there are other elements that often combine to create droughts. Heat waves often intensify drought conditions by taxing ground water in the soil as well as surface supplies. Rain-producing wind patterns may also shift, taking rain-laden clouds elsewhere. Atmospheric high pressures and changing ocean currents may also aid in producing drought conditions. Sunspots have also been cited as affecting rainfall and contributing to drought.

Measurement

Droughts have a variety of effects on man, and on the world's plant and animal life. The seriousness of a drought depends on its intensity and length. A drought becomes serious once vegetation and crops wilt and surplus water supplies in wells and surface reservoirs are overtaxed or depleted. In today's technological society we use more water per person than did previous generations, and often impair replenishment of ground water by filling wet lands and aquifer recharge areas.

The extent of the drought may be measured by the drawn down level of an open water supply such as a reservoir or by the drop in the ground water level. Drought intensity can also be measured by its impact on farming, irrigation, municipal and industrial water supply, power generation, navigation, and recreation. Water supply problems are monitored by state and local officials. Emergency measures such as resource sharing sometimes become necessary if the problem becomes severe.

Secondary Effects

Fires in urban areas can be aggravated because of inadequate water supplies or reduced water pressure in water mains—the result of community water shortage or drought. Parched forests are especially vulnerable to serious, raging fires, whether ignited by man or by lightning.

Detection

Observers have attempted to anticipate drought cycles, by examination of tree rings, lake levels, glacier flows, and human and animal migrations from the past. Existing droughts can be determined by declining crop production that may occur steadily over a number of years. A decline in water storage supplies caused by consumption and insufficient water replacement is a primary signal that a drought is at hand. Analysis of rainfall, runoff, and weather conditions compared to the average precipitation over a period of time, will clearly show conditions of drought. Numerous studies have been completed which outline potential community water supply problems; which can be severely aggravated by the underestimation of future water needs, poor plant maintenance, pollution, and contamination.

FLOOD

Nature and Cause

A flood is an unusually heavy overflow of water from streams or other water bodies that spreads out over fields, cities, and towns of adjacent flood plains, causing damage, destruction, and death in the water's downhill surge toward the sea.

A flood is caused by the inability of excess water to be taken up by the soil, vegetation, or atmosphere or the inability to be drained off in water channels or stored in water retention facilities such as lakes and reservoirs. Inland floods can be caused by excessive rainfall, dam failure, landslides, volcanoes, or rapid snow melt. Coastal floods can be caused by hurricanes, blizzards, or any severe ocean storms that creates tidal surges that push sea waves onto low-lying coastal areas. The most serious coastal flooding is cyclical, usually occurring in the spring when abnormally high tides are accompanied by strong winds.

Measurement

The severity of a flood is defined either in terms of the maximum height of the water above flood stages, or as the maximum amount of water flowing past a certain point. Height is recorded in linear feet and is easier for the public to understand than the water flow or discharge rate, which is measured in cubic feet per second. The discharge rate is the most accurate basis for comparing floods occurring in the same place at different times.

Secondary Effects

These may include dam/levee failures, transportation accidents, landslides, power failures and water supply contamination/failure. Floods may also cause considerable indirect damage from ill health, disease, and lost work time to businesses. Flood-caused soil erosion, which results in the down stream settlement of silt and sediment, causes damage by clogging reservoirs and destroying fertility of farmlands.

Detection

Flood detection occurs through the utilization of several systems that monitor the input of water into rivers, streams, and coastal lowlands. A constant watch is kept over weather, river, and soil conditions at all large rivers and throughout river basins. The Army Corps of Engineers oversees a vast system of flood-control works and maintains a flood control reservoir reporting system.

HURRICANES

Nature and Cause

A hurricane is a violent wind and rain storm characterized by a counterclockwise spiral of wind that covers an area of up to 300 miles in diameter and generates wind speeds of 74 to 200 miles per hour. The highest wind speeds occur in the band around the storm's center, or "eye." The area within the eye of the storm and the outer periphery are relatively calm, with winds seldom exceeding 30 miles per hour. A hurricane can dominate tens of thousands of square miles of the earth's surface and bring with it grave danger. A hurricane is most violent as it travels over the ocean. After reaching land, the wind speed will be reduced, but the area vulnerable to damage becomes much broader.

Hurricanes usually occur between June and November and originate in the Caribbean or the Gulf of Mexico, or sometimes as far east as the Lesser Antilles or the Cape Verde Islands. A hurricane begins as a tropical cyclone and grows to immense proportions. It is formed when a polar air mass hits an area of convection at the warm sea surface. The warm air core ascends, causing heavy precipitation and violent winds.

Measurement

The National Weather Service uses the Saffir/Simpson Scale to measure a hurricane's severity. This scale divides hurricanes into five categories:

Minimal. No real damage to building structures. Damage primarily to unanchored mobile homes, shrubbery, and trees. Also some coastal road flooding and minor pier damage. Winds range from 74 to 95 miles per hour and/or tidal surges reach 4–5 feet above normal.

Moderate. Some damage to roofing material, doors, and windows. Considerable damage to vegetation, exposed mobile homes, and piers. Coastal and low-lying escape routes flood two to four hours before arrival of storm center. Small craft in unprotected anchorages break moorings. Winds range from 96 to 110 miles per hour and/or tidal surges reach 6–8 feet above normal.

Extensive. Some structural damage to small residences and utility buildings, with a minor amount of curtain-wall failures. Mobile homes are destroyed. Flooding near the coast destroys smaller structures, and larger structures are damaged by floating debris. Terrain continuously lower than five feet may be flooded inland six miles or more. Winds range from 111 to 130 miles per hour and/or tidal surges reach 9–12 feet above normal.

Extreme. More extensive curtain-wall failures with some complete roof-structure failure on small residences. Major erosion of beach areas. Major damage to lower floors of structures near the shore. Terrain flooded inland as far as eight miles. Winds range from 131 to 155 miles per hour and/or tidal surges reach 13–18 feet above normal.

Catastrophic. Complete roof failure on many residences and industrial buildings. Some complete building failures with small utility buildings blown over or away. Major damage to lower floors of all structures located less than 15 feet above sea level and within 500 yards of the shoreline. Wind speeds exceed 155 miles per hour and/or tidal surges exceed 18 feet above normal.

Secondary Effects

The most threatening elements that accompany a hurricane are the "storm surge" and the flooding. The storm surge is a great dome of water that travels with the hurricane and breaks on the shore, usually destroying everything in its path. Flooding that results from this major surge and attendant secondary surges is the prime cause of hurricane-related deaths and property damage. In inland areas, heavy rainfalls bring about flash floods. Wind damage to property is not as significant as flooding damage, but average winds are usually around 80 miles per hour and can measurably contribute to the storm's devastating effects. Occasionally, tornadoes will be generated by hurricanes.

Detection

The National Weather Service detects and tracks hurricanes through the Hurricane Forecast Center in Miami, and reports their progress to the public through the Severe Storm Warning Center in Kansas City. With the aid of a radar fence that reaches from Texas to New England, forecasts are issued every six hours under normal circumstances, and every three hours when a hurricane watch or warning is in effect.

A hurricane becomes a threat to the United States when its somewhat erratic path moves northwesterly up the Gulf or Atlantic coasts. A hurricane that affects thousands of square miles can be one of the most destructive forces on earth.

TORNADO

Nature and Cause

A typical tornado is a swirling storm of short duration, with winds of up to 300 miles per hour, having a near vacuum at its center. It appears as a rotating funnel-shaped cloud, from gray to black in color, extending towards the ground from the base of a thundercloud. Tornadoes normally cover a very limited geographical area and usually give off a roaring sound. A tornado is the most concentrated and the most destructive form of weather phenomena.

Tornadoes are usually the result of the interaction of a warm, moist air mass with a cool or cold air mass. They often occur out of squall lines and thunderstorms, particularly in the afternoon. Nocturnal occurrences are infrequent but have been recorded. Tornadoes may occur in any season but are more prevalent in spring and summer.

Measurement

The usual measurement of a tornado is the estimated wind speed at the core, degree of damage to structures, and the number of persons injured or killed. Often, the total dollar cost of rehabilitating tornado-damaged structures is used as a measuring criterion.

Secondary Effects

Usually, flash flooding, electric power outages, transportation-system and communication-system disruption, and fires are the most common secondary effects of tornadoes.

Detection

Whenever weather conditions develop that may indicate that tornadoes are expected, the National Weather Service will issue a tornado watch to alert people in a designated area for a specific time period (usually six hours). The tornado watch is upgraded to a warning when a funnel cloud is actually sighted.

EARTHQUAKE

Nature and Cause

Earthquakes are caused by the release of energy in the earth's crust. The released energy moves masses of earth in sudden, shifting motions ranging from slight tremors to violent, destructive shocks. Dislocations along the earth's crust are produced by pressure generated deep inside the earth. The earth's crust responds to this pressure by bending and eventually snapping or shearing. When the crust snaps, seismic waves are sent through the crust outward from the center of the quake. These waves may travel long distances at varying speed, depending on the force of the quake and the composition of the earth's surface.

When earthquakes do occur, the seismic waves may have a devastating impact on anything within their paths, such as buildings, roads, and bridges, causing losses in both lives and properties. The degree of damage depends on the force of the seismic shock waves, structural integrity of the buildings, and the composition of soils and rock that underlie the affected area.

Measurement

Seismic waves or vibrations that cause earthquakes are detected, recorded, and measured on instruments called seismographs. Seismographs are highly sensitive monitors that can pick up earthquake tremors at great distances. Seismographs use the Richter scale to measure the degree of magnitude or intensity of a recorded earthquake. The Richter scale is a numerical, logarithmic system where a 3-scale quake is 10 times as powerful as a 2-scale event.

Also used is the modified Mercalli scale, measuring earthquake intensity at a particular location, on a scale form I to XII, based on observed effects. The Richter scale is more widely used, paired with seismographic stations throughout the world.

Secondary Effects

In addition to the direct destruction of buildings caused by earthquake shocks, quakes may trigger devastating secondary hazards such as landslides, avalanches, dam bursts, fuel-line fires, tidal waves, utility outages, water-supply interruption, road and rail accidents, chemical spills, and pollution.

Detection

Seismographs are located throughout the country for recording the trends and directions of even the smallest earthquakes.

WINTER STORM

Nature and Cause

There are two distinct types of winter storms: ice storms and snowstorms. Ice storms are characterized by freezing rain that forms a layer of ice on roads, trees, and other objects. Ice storms occur when surface temperatures are below freezing and the liquid precipitation freezes upon impact. Rain that freezes *before* reaching the ground is known as sleet.

Snowstorms are characterized by a moderate to heavy fall of snow often accompanied by high winds. Snowstorms are caused by the same type of atmospheric conditions that bring about summer thunderstorms, but, because the temperature is cold, the precipitation falls as snow instead of rain. Masses of polar and tropical air confront each other, bringing about intense low-pressure systems that can churn over areas as great as tens of thousands of square miles.

Measurement

Both ice storms and snowstorms are measured by the severity of their effects, amounts of precipitation, and various other meteorological factors. A mild freezing rain is short in duration and leaves ice coatings of less than an inch. A severe freezing rain can last for hours and leave objects covered with an ice layer of up to eight inches.

In most areas, accumulation of six inches of snow is not considered a major hazard, and little official note is made of a snowstorm until and unless it reaches blizzard proportions, characterized by heavy snowfall (sometimes accumulating to a foot or more a day over several days), winds of at least 35 miles per hour, and temperatures of 20° Fahrenheit or lower.

Secondary Effects

Ice storms glaze roads and bring about many traffic-related injuries and deaths. Thick layers of ice on trees limbs and utility cables cause them to collapse, bringing about injuries, property damage, and frequently, power outages of disastrous proportions. Weather-related effects of coastal snowstorms are tidal surges from high winds and subsequent flooding.

Victims of both ice storms and snowstorms fall into these categories: One-third of winter-storm deaths are caused by automobile and other accidents; another one-third are attributed to overexertion and resulting heart attacks; one-tenth are due to overexposure and fatal freezing; and the remainder to a variety of causes including home fires, carbon-monoxide poisoning in stalled cars, falls on ice, electrocution from downed wires, and building collapse.

Detection

Winter storms are detected, monitored, and reported by the National Weather Service through the usual weather and media networks. The terms "watch" and "warning" are used for winter storms as for other weather related threats. The watch alerts the public that a storm has formed and is approaching the area. A warning is issued when 6 inches or more of snow is expected within the next 12 hours. This is increased to "severe warning" status when, in addition to the predicted snow, the temperature drops to 10° Fahrenheit or lower and/or the wind speed increases to 45 miles per hour.

3

Technological Hazards

Whereas the occurrence of natural disasters does not experience radical shifts in frequency, the frequency of technological events is ever increasing, and the growing potential for technological disasters is dramatic. It appears that the overall management of natural emergencies is more coordinated and less fragmented than is the management of technological hazards. Surveys indicate that technological events entail:

- less warning
- shorter duration
- less response coordination
- less federal involvement
- greater local and private involvement

NUCLEAR POWER PLANT FAILURE

Nature and Cause

Nuclear power plant failure poses a threat to the population within a 10-mile radius of the site—and in very severe cases within 50 miles of the site—when a breakdown in the plant's systems occurs. The severity of the threat is in direct proportion to the degree of uncontrolled activity within the reactor and the amount of radiation that subsequently escapes. Nuclear power plant incidents are caused by operational error or system malfunction. Some possible causes of nuclear power plant incidents are sabotage, natural disasters such as earthquakes, and the possible leakage of radiation from spent fuel that is stored on site.

Measurement

The Nuclear Regulatory Commission (NRC) has classified nuclear power plant incidents into four categories, from least to most severe:

Unusual Events. Events are in process of occurring, or have occurred, and indicate a potential degradation of the level of safety of the plant. No releases of radioactive material requiring off-site response or monitoring are expected unless further degradation of safety systems occurs.

Alert. Events are in process of occurring, or have occurred, and involve an actual or potential substantial degradation of the level of safety of the plant. Any releases expected to be limited to small fractions of EPA protective-action-guideline exposure levels.

Site Area Emergency. Events are in process of occurring, or have occurred, and involve actual or likely major failures of plant functions needed for protection of the public. Any releases are not expected to exceed EPA protective-action-guideline exposure levels except within site boundary.

General Emergency. Events are in process of occurring, or have occurred, and involve actual or imminent substantial core degradation or melting with potential for loss of containment integrity. Releases can be reasonably expected to exceed EPA protective-action-guideline exposure levels off-site for more than the immediate site area.

Secondary Effects

Leakage of radiation from a nuclear power plant causes radioactive pollution of the air and, under more severe circumstances, ingestion of radioactive material into the water and soil.

Detection

Nuclear power plants have built-in safety monitoring systems that will trigger an initial warning when something goes awry. From that point, emergency response plans insure that proper public officials will be notified of the incident. Agencies included in the notification system are state police, state departments of health, and state civil defense agencies, who in turn, coordinate necessary emergency response procedures.

HAZARDOUS MATERIALS ACCIDENT

Nature and Cause

A hazardous materials accident is an unexpected turn of events that brings about spilling, leakage, or other mishap involving a dangerous substance. These accidents occur in the manufacturing, storage, use, and transport of the substances. Several thousand potentially dangerous chemicals are in daily use, and their mishandling can cause emergency situations that create an environment of varying degrees of danger to significant numbers of people.

Although accidents involving hazardous materials can occur anywhere that the substances happen to be present, the majority occur on major highways. Hazardous materials fall into one or more of these general categories:

1. **Explosive liquids and gases**, such as propane.

2. **Flammable liquids**, such as alcohol.

3. **Caustic substances**, such as chlorine which can cause breathing problems.

4. **Disease-carrying substances**, such as medical serums.

5. **Corrosive materials**, both liquid and solid, which can destroy skin tissue.

6. **Radioactive materials.**

7. **Poisons**, which, through ingestion, absorption, or inhalation, can bring about immediate or delayed infirmity or death.

All such chemicals carry fumes that are extremely harmful when inhaled, absorbed, or ingested. Some are deadly poison, but most are flammable and, under certain conditions, explosive. All are potentially life-threatening if leaked into water supplies. Some react violently on contact with water, and many have other deadly characteristics such as causing suffocation or destroying skin tissue.

Hazardous materials accidents are usually caused by human error. Substance handlers usually misjudge or miscalculate because of unfamiliarity with the product. Accidents also occur when the equipment related to the use or transportation of a substance malfunctions or fails. In addition, some accidents result from lack of, or loose enforcement of, laws regulating hazardous materials, or by intentional breaking of those laws.

Measurement

The effects of hazardous materials accidents can readily be measured in terms of injuries, deaths, property damage, and the millions of dollars they cost each year. Although all hazardous substances are potentially dangerous, some carry a far greater threat to life and property than do others.

Secondary Effects

The spilling or leakage of dangerous chemicals can bring about major explosions, fires, penetration of toxic fumes into the air, immediate and long-range poisoning of people and animals, and pollution of water and soil. These occurrences, in turn, often necessitate the evacuation of large numbers of people from their homes and businesses.

Detection

Hazardous materials accidents are usually detected by the effects they bring about. Emergency personnel and area residents can readily perceive a fire, an explosion, or toxic fumes. Occurrences of these accidents are reported through the usual emergency communications and media networks. Other organizations such as CHEMTREC and the EPA have nationwide hotline numbers that provide information for handling the spill or leakage. Local branches of these organizations dispatch investigators and advisors to the site of the accident.

CONFLAGRATION

Nature and Cause

A conflagration is a large, destructive fire, and when occurring in a closely compacted residential or business district, it spreads rapidly and is extremely difficult to control. There are four classes of fires as determined by the National Fire Data Center of the Federal Emergency Management Agency, which, when taking place individually or combined, can cause major urban fires:

Class A, caused by fuels that leave glowing coals.

Class B, caused by fuels that leave no glowing coals.

Class C, caused by dangerously high voltage and/or faulty wiring.

Class D, caused by combustible metals, incendiary materials, or other explosive type devices.

Fires that are grouped by class according to cause; however, a fire can exist only when the proper proportions of heat, oxygen, and fuel are combined.

Measurement

Conflagrations are statistically measured by number of incidents and deaths, and by cost of damages.

Secondary Effects

Major urban fires may cause other disasters including power failures, water supply problems, and hazardous materials accidents. There could also be air pollution; loss of housing, jobs, and revenue; and other long-term economic problems.

Detection

Conflagrations will usually be detected as a result of causative factors affecting a fire within a specific area. These fires may be detected through a series of warning devices located inside or outside public and private buildings. Fire alarms and smoke detectors installed within a building are parts of an internal system to alert occupants. Fire alarms—owned, maintained, and centrally located by the municipality—are normally in public areas. These alarms are connected to a central fire dispatcher, and when activated cause an immediate firefighting response. The recognition of intense heat, flames, or smoke by individuals is another detection method that will cause firefighting personnel to be notified.

MAJOR TRANSPORTATION ACCIDENT

Nature and Cause

A major transportation accident is an unexpected incident involving any means of human transport resulting in a situation beyond the control of local public safety personnel. Transportation accidents may occur as the result of human error, design or maintenance flaw in the vehicle or system, sabotage, or terrorism. Faulty operator judgment is the root cause of most transportation accidents, although it may be in combination with other factors existing at that time.

Measurement

Transportation accidents are usually measured by the number and severity of the injuries inflicted and by the number of deaths. The number of vehicles involved, the destruction or damage caused to public and private property, and the inconvenience to those using the system are other possible measures as well. Often, system repair, replacement costs, and the medical-care costs of the injured are also used as measurements.

Secondary Effects

Among many possible secondary effects are fire, explosion, hazardous-material spill, panic, looting, system disruption, system breakdown, and system boycott by potential users.

Detection

Missing vehicle reports, reports by public safety personnel, reports from survivors, and news media reports are the most likely detection methods.

DAM FAILURE

Nature and Cause

Dam failure is a hazard not as easily categorized as flood or nuclear power plant failure. Dams are highly vulnerable to natural forces, but, as man-made structures, they are continually dependent on the quality of their construction and maintenance. Therefore, dam failure is included here, as a technological hazard.

Dam failure is defined as the catastrophic giving way of a dam, characterized by the sudden, rapid, and uncontrolled release of impounded water. There are lesser degrees of failure, and any malfunction or abnormality that adversely affects a dam's primary function of impounding water is properly considered a failure. Any lesser degree of failure can progressively lead to or heighten the risk of a catastrophic failure.

Dam failures can be caused by excessive rainfall, runoff increased by rapid snow melt, poor construction or maintenance, earthquake, or various other natural or man-caused phenomena.

Measurement

Dam failures are measured in terms of potential flood threat to downstream property and inhabitants. This flooding potential is depicted on an inundation map. The map shows the outlines of the anticipated flood area in enough detail to identify dwellings and other features that are expected to be flooded by the dam break. Information on depth of flooding and estimated time of downstream travel to specific locations are also used to measure the severity of dam failures.

Secondary Effects

Dam failures are likely to cause many other disasters, particularly water-related ones. A flash flood and/or a slow-rise flood are the most prevalent effects and often lead to additional downstream dam/levee failures. Water supply problems, power failure, release of hazardous materials, and even radiological incidents can also be triggered by dam failure.

Detection

Detection prior to dam failure is dependent upon an established inspection schedule and a timely warning system. The dam-inspection program will verify the structural integrity of the dam and appurtenant structures. Inspections will normally disclose unsafe conditions in time for corrective action to be taken. The dam owner should establish an effective monitoring system when weather and other pertinent data indicate the need. Such a system will monitor the progression of potential dam failure and report estimated time available for evacuation and other protective actions.

FOREST FIRE

Forest fire is another hazard not easily categorized. Although countless catastrophic forest fires have been the direct result of natural forces, man's role as an active contributor to the problem is significant enough to include forest fire as, at least partially, a technological hazard.

Nature and Cause

A forest fire is the uncontrolled burning of sparsely populated areas covered by vegetation or forest cover. Forest fires may be caused by a multitude of triggering events, such as lightning, man's carelessness with smoking materials and campfires, pyromania, and spontaneous combustion.

Forest fires are usually measured in terms of the fire front advance per unit of time or in terms of the acres consumed per unit of time. The monetary loss to structure, timber, and other forest products, and the dollar cost of reforestation and rehabilitation are measures as well.

Secondary Effects

The destruction of timberland and wildlife habitat; injury and death to humans, animals, and birds; plus the possibility of erosion and mud slides in the land denuded of ground cover are some of the possible secondary effects.

Detection

Public safety personnel reports (state and federal forest monitors), orbiting satellite reports, aircraft crewmen reports, and information from forest users are means of detecting forest fires.

4

Developing Emergency Plans

AN APPROACH TO EMERGENCY PLANNING

There are several approaches to emergency planning that will result in an effective emergency plan. One approach that is sometimes promoted by well-meaning individuals, but is not recommended, is the "fill-in-the-blank" approach. This method often results in an ineffective plan because it assumes that certain response capabilities exist that in fact do not. This type of plan is sometimes referred to as a "paper tiger" plan, as it appears to be a strong and effective plan on paper but in reality cannot be implemented effectively. The plan should be based on correct assumptions and actual capabilities. The only way to ensure this is to use a process that involves many facility personnel in the planning. It is important to bring together those individuals whose joint responsibility will be the management and execution of emergency operations.

Dwight D. Eisenhower hailed the value of the planning process when he said, "Plans are worthless, but planning is everything." Perhaps a more appropriate message for this book is "planning is vital, but plans are the source of actions." Plans reflect the decisions made during the planning process. The plan becomes the blueprint for emergency procedures and will be used and referred to frequently during training for new employees, refresher training for existing employees and managers, and as a reference tool under certain emergency circumstances.

It is recommended that a planning group be established to coordinate the development of the emergency plan. While this group does not have to be involved in every aspect of emergency plan development, it should provide overall direction and review the progress of the process. A manager or an outside consultant can do the basic work and the actual writing of the plan, if so desired, but regular meetings of the planning group are important so that continuity of effort can be maintained. Another approach is to have members of the group write individual sections of the plan and then combine and edit the plan as a whole.

It is important to note that organizations outside the company should be involved as well. For example, if a chemical plant is developing a new plan, the local fire and police departments, among others, should be involved.

STEPS IN EMERGENCY PLANNING

There are a number of steps to be followed in developing an emergency plan. These are illustrated in figure 4–1.

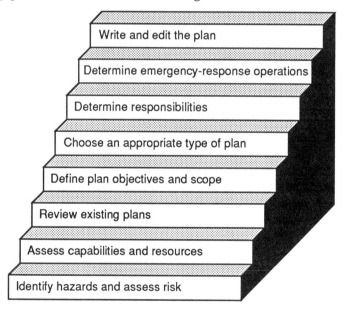

Writing the emergency plan is the last step in the plan development process

Figure 4–1: Steps in emergency planning

Step 1: Identify Hazards and Assess Risks

The development of your preparedness program must begin with a thorough knowledge of potential hazards that could adversely affect your business. This function is known as risk assessment. Risk assessment is a natural function that is performed by everyone—everyday. Whether crossing the street or driving down a highway, one continually looks for danger by analyzing conditions that are present and then attempts to visualize the various things that could happen.

A better explanation of risk assessment might be to describe it as the process of determining the adverse consequences that may result from the use of a technology or from some other action. The assessment of risk typically includes (1) an estimate of the probability of the hazard occurring, (2) a determination of the types of hazards posed, and (3) an estimate of the number of people or things likely to be exposed to the hazard as well as the number likely to suffer adverse consequences.

Effective risk assessment in an industrial facility will help prevent most emergencies because most dangerous conditions will be recognized and corrected. In addition to its prevention characteristics, the value of this step is that it creates an awareness of what should be planned for and the extent of impact a hazard will pose. It will therefore help identify what sort of commitment or help your facility will require in combating the emergency. When rare emergencies do occur, you will be ready and able to respond more effectively.

TECHNIQUES

The following section outlines the various methods available to identify hazards and assess risk. *It is not intended to provide instruction on performing these techniques.*

A facility's vulnerability to natural hazards can be determined in a variety of ways. The simplest and least expensive method is to check with your local emergency preparedness office (sometimes referred to as the emergency services office, civil defense agency, or emergency management agency) and obtain a copy of the Hazard Analysis Study for your community. This qualitative study, prepared under the guidance of the Federal Emergency Management Agency in many U.S. communities, outlines the history, vulnerability, and probability of various natural (and some man-made, or technological) hazards in your community. Even if such a study is not available, the director of your local agency will likely have a good idea of the risks to the community and should be able to provide you with some basic information.

You should also identify the hazards likely to be present at your facility due to the nature of chemicals that may be stored, processes used to produce materials, or similar hazard-related factors.

Qualitative Techniques

A **hazards and operability study** (HAZOP) is a means to determine the possible causes and consequences of abnormal process conditions such as equipment failure and human error. The HAZOP is a process of a hazard identification that is performed according to certain standard techniques, some of which are described in the *Guidelines for Hazard Evaluation Procedures* published by the American Institute of Chemical Engineers' Center for Chemical Process Safety. Essentially, this study is a brainstorming session to identify possible accident scenarios.

A **failure mode and effects analysis** (FMEA) is used to evaluate hazard potential of any system by considering the effects of each of the possible failure modes of each component and its affect on the entire system. The various failure possibilities and effects are tabulated and each is ranked according to its criticality.

A **process safety audit** (PSA) is an inquiry into potential process-related hazards that is performed by an on-site auditing team. The team thoroughly evaluates such items as operating procedures and record keeping, process modification procedures, pre-emergency and response planning, and personnel training. If desirable, maintenance procedures and record keeping, equipment testing procedures, and the basis and adequacy of process-safety protective systems are also examined.

The **Dow and Mond Index** provides a method for relative ranking of the risks in a chemical process plant. The method assigns penalties to process materials and conditions that can contribute to an accident. Credits are assigned to plant safety procedures that can mitigate the effects of an accident.

Quantitative Techniques

A **hazard analysis** is a method of quantifying the frequency of occurrence of accident scenarios and the magnitude of the associated consequences. The methods used are generally those of the **fault tree** and/or **event tree**. Probability calculations are based on many sources of plant-specific information including design, operating, and maintenance procedures; historical failure-rate data; and plant operator experience. The impact of the accident on the surrounding population is then determined by predicting the dispersion of hazardous material and taking into account its toxic or flammable properties. A blend of both fault-tree and event-tree analysis is referred to as **Cause-Consequences Analysis**.

Air dispersion modeling is a method of analyzing the characteristics of the release of a hazardous material. An example of such modeling is the ability to predict the rate of evaporation from a pool of spilled liquid, the atmospheric dispersion of heavier-than-air vapors, the interaction of toxic clouds with the surrounding population, and the effects of vapor-cloud explosions and fires.

Figure 4–2 illustrates the primary and secondary purposes of a variety of hazard evaluation techniques and is taken from the EPA's "Review of Emergency Systems Report to Congress."

Step 2: Assess Capabilities and Resources

Upon completing a hazard analysis, you will be aware of the magnitude of the problem(s) you could face during an emergency. Therefore, you should be thinking of the types of resources that will be needed to ensure an effective emergency response capability. Resources that you may need for an effective response may be different from resources that are currently available, so you must perform an assessment of current resources and capabilities. Resources include facilities, equipment, and supplies, while capabilities may include the staff expertise, experience, or training. If deficiencies are identified, you should plan to correct them as soon as possible. Remember, the emergency plan should be written based on only *current* resources and capabilities.

The following questions may be useful in assessing resources and capabilities.

1. Does the facility have adequate on-site emergency response equipment such as for firefighting, personal protection, and communications?

2. Has an inventory of equipment been prepared?

3. Are trained personnel available to provide on-site emergency response such as firefighting, or spill control?

4. Does the plant have medical/first-aid capabilities?

5. Have arrangements been made with local hospitals for providing treatment for chemical exposures and other medical emergencies?

6. Are employee emergency evacuation plans in place?

7. Have employees been trained to use emergency evacuation plans?

8. What on-site notification systems are available?

9. What on-site communications systems are available?

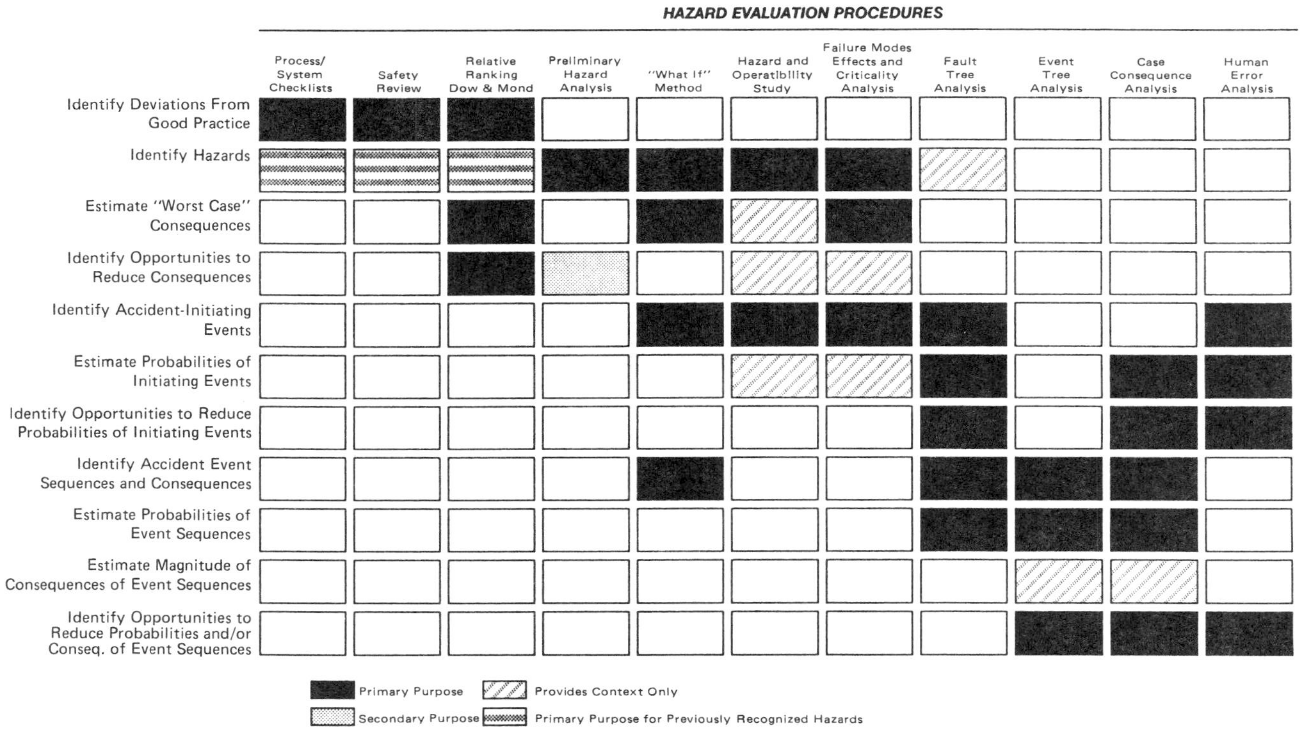

HAZARD EVALUATION PROCEDURES

Columns (Hazard Evaluation Procedures):
- Process/System Checklists
- Safety Review
- Relative Ranking Dow & Mond
- Preliminary Hazard Analysis
- "What If" Method
- Hazard and Operability Study
- Failure Modes Effects and Criticality Analysis
- Fault Tree Analysis
- Event Tree Analysis
- Case Consequence Analysis
- Human Error Analysis

Rows:
- Identify Deviations From Good Practice
- Identify Hazards
- Estimate "Worst Case" Consequences
- Identify Opportunities to Reduce Consequences
- Identify Accident-Initiating Events
- Estimate Probabilities of Initiating Events
- Identify Opportunities to Reduce Probabilities of Initiating Events
- Identify Accident Event Sequences and Consequences
- Estimate Probabilities of Event Sequences
- Estimate Magnitude of Consequences of Event Sequences
- Identify Opportunities to Reduce Probabilities and/or Conseq. of Event Sequences

Legend:
- ■ Primary Purpose
- ▨ Primary Purpose for Previously Recognized Hazards
- ▥ Secondary Purpose
- ▨ Provides Context Only

Source: The Center for Chemical Process Safety, American Institute of Chemical Engineers. Guidelines for Hazard Evaluation Procedures. Prepared by Battelle Columbus Division. New York, New York, 1985.

Figure 4–2: Hazard evaluation procedures

10. What systems are available for plant/community communications?

11. What systems are available to warn neighbors of toxic releases?

12. Has the public been educated to know what to do in case it hears alarms from the plant?

13. How do site representatives coordinate with local government officials during an emergency?

14. Does the plant belong to an industrial mutual aid association, or are neighboring facilities' resources available during an emergency, if needed?

15. If yes, what resources?

16. Does the facility have contracts with spill-cleanup specialists?

17. Does the facility have in-house capability for spill cleanup?

18. How does the site determine concentrations of released chemicals?

19. Are toxic-gas detectors, explosimeters, or other detection devices positioned around the site? Where?

20. Does the facility have meteorological equipment (such as for measuring wind direction or wind speed) installed?

21. Does the facility have a capability for modeling vapor-cloud dispersion?

22. Does the facility have auxiliary power systems available to support emergency functions in case of power outages?

23. Is there a program for periodically inspecting emergency equipment and supplies? Is the program regularly followed?

While the preceding list of questions is comprehensive, it is not all-inclusive and therefore should be expanded and modified to meet your facility's needs. The list is provided only to demonstrate the types of questions that should be considered during this assessment. It is important to note that resources and capabilities should be assessed in relation to expected hazards. For example, if your facility does not use chemicals, the questions concerning dispersion modeling are unnecessary. But if your facility is in a flood plain, questions concerning the availability of pumps and sandbags would be appropriate. Similar questions should be asked about the community's capabilities and resources.

Step 3: Review Existing Plans

Before you "reinvent the wheel," it would be wise to check old files to determine whether your facility ever had an emergency plan. If so, updating the old plan may be all this is required—with a word of caution, however: An old plan was written for hazards that existed at the time of its publication, and since the business and its associated hazards have grown and changed, an old plan may no longer be appropriate. Other things that change with time are local capabilities, equipment and processes in use, and regulatory requirements. So while reviewing an older plan may be helpful, it should not be considered a substitute for devising a new and up-to-date emergency plan.

Another good idea is to review neighboring facilities' plans in order to understand how they prepare for emergencies. Many organizations have emergency plans that might be of use in the development of your facility's plan. Their plans may provide insight as to how your plant emergency brigade should interface with community response agencies. Caution should again be exercised so that the mistakes or deficiencies of another facility's plan do not find their way into your own plan.

Next, review your community's emergency plan(s). The local civil defense or emergency preparedness office is the most likely source for this information. With the advent of Local Emergency Planning Committees (LEPCs), communities are in the process of developing new hazardous materials response plans. By reviewing such a plan, you should be able to determine how your response activities would be affected by local response actions. In addition to the community plan, you should also review the fire and police departments' emergency operations procedures. These procedures outline exactly how each department will carry out its emergency responsibilities. State and federal agencies may also have plans that could impact your own operations, such as the National Oil and Hazardous Substance Pollution Contingency Plan or the EPA's regional response plans. In addition to these emergency plans and procedures, other documents may contain important information that would be of use in developing your own plan, such as government mutual-aid plans, agency rosters, and local ordinances.

Step 4: Define Plan Objectives and Scope

At this point, hazards, resources, and capabilities have been assessed and existing plans have been reviewed. The next step is to identify and develop the objectives for your emergency plan. In other words, what do your hope to accomplish by having an emergency plan? For example, do you want the plan to outline emergency response actions only, or do you want it to cover prevention and recovery actions as well? What hazards will the plan cover? By defining these objectives now, your plan can be focused and concise.

Step 5: Choose an Appropriate Type of Plan

Once steps 1–4 have been completed, you will be able to choose the type of plan that best suits your needs and objectives. While there is widespread disagreement between experts concerning semantics, there are essentially four basic types of plans commonly used in industry: (1) actions guides or checklists, (2) response plans, (3) emergency management plans, and (4) mutual aid plans. Generally speaking, every plan fits into one of those types, and it is up to you to decide which type is best for your situation. The level of comprehensiveness and sophistication of plans in each category will vary depending on your facility's particular needs.

An **action guide or checklist** is generally a short and simple way to describe basic procedures that must be followed, such as whom to call, information to be collected, and basic response functions. It is intended to be more of a reminder as opposed to being a comprehensive plan, and should not, therefore, be used by an untrained individual. A guide or checklist is sometimes used to augment another type of plan.

A **response plan** is usually very detailed and designates responsible individuals and the actions they must implement to mitigate the problem at hand. Sometimes a response plan is written for each type of possible hazard at the site (which is also know as contingency planning). For example, there might be a chemical-spill response plan as well as a hurricane response plan. Some response plans deal with all possible types of hazards and specify functions that certain individuals are to perform regardless of the hazard. Response plans deal only with actions necessary to mitigate an emergency and do not cover requirements for prior preparations (e.g., training, drills, and exercises) or for postdisaster concerns (e.g., recovery plans and business interruption plans).

An **emergency management plan** is the most comprehensive type of plan used in business. It usually states who does what, when, and how—before, during, and after a disaster. Often, this type of plan will incorporate implementing procedures that detail how a particular responsibility is to be carried out. The benefit of this type of plan is its comprehensiveness. It deals with emergencies in four logical phases: prevention, preparedness, response, and recovery.

Finally, some companies will develop with nearby firms a **mutual aid plan** that calls for the participating firms to share resources and to help one another during an emergency. This type of plan demands a lot of coordination to implement, but can be very useful for small firms with limited resources or for larger firms with high hazard potential.

Step 6: Determine Responsibilities

One of the key elements of an emergency plan is that it states who is responsible for doing what. Therefore, before writing the plan, you must decide organizational responsibilities, for emergency operations. To do this, you should use an organizational chart. You will need to compile a list of key emergency functions that need to be addressed in the plan (refer to table 5–1 on page 42). Set up a matrix with individuals/departments on one axis and emergency functions along the other. Then select those who will have primary and secondary responsibilities for each emergency function. When making assignments, remember to consider the experience, expertise, and manpower available to accomplish each function. For example, site security and crowd control are the logical responsibilities of security guards; yet if there are only two security guards on duty during back shifts (nights, weekends, or holidays) they cannot be expected to carry out these functions without additional support. All emergency assignments should be reviewed with the plant's emergency planning group.

An emergency response organization will probably be different from the normal plan organization. Therefore, an organizational chart specifically for emergencies needs to be developed with the emergency plan and included for all to see, along with the responsibility matrix. This will help clarify emergency authority. Also, be sure to establish a line of succession at least three-people deep for all primary emergency positions.

Step 7: Determine Emergency-Response Operations

At this point, you will have identified who is to do what under certain circumstances, but have not yet determined how they are to do it. And while you have identified major responsibilities, you have not yet identified specific tasks that are to be performed in carrying out those responsibilities.

Defining the concept of operations is a critical step because it lets you map out how the plant's emergency response will really work. One of the best methods to do this is to use a flow chart to identify all the activities that will take place from the moment an emergency arises. By developing and refining this flow chart, you will end up with a graphic display of the concept of operations. The flow chart will help you identify the management system that will be utilized in time of emergency and to identify deficiencies or gaps in your thinking. Once this chart has been developed and completely reviewed and critiqued, you will have a good idea of what must be included in the plan.

Step 8: Write and Edit the Plan

Steps 1–7 of the emergency planning process having been completed, the last step is simply to put the plan on paper in an orderly, concise, and complete manner. It is best to limit actual writing to one or two people to help ensure consistency in style and format. Drafts of the plan should be reviewed for technical merit by the emergency planning committee so that nothing that should be included is omitted. Once the plan is deemed to be technically complete, it is a good idea to have a professional editor review the plan to ensure that is it written correctly and clearly. Before the final printing, all edited matter must be reviewed for technical accuracy as important details can often be lost or misinterpreted during an edit.

COMMON PITFALLS IN EMERGENCY PLANNING

The following are some of the more common reasons given for the breakdown of the emergency planning process.

Plan developed by an individual. A plan developed by a single individual, or by a few individuals, is sometimes in danger of not being complete. Also, the plan may be unknowingly slanted or biased. These problems can be avoided if the plan is thoroughly reviewed by larger groups of people, more broadly representative of the entire facility.

Plan developed by a group. While groups will most likely ensure a comprehensive planning effort, group planning can have some drawbacks. Groups tend to work at a relatively slow pace. Meetings can be long-winded and unproductive. Finally, interorganizational problems often surface during the group planning process. A strong group chairman can often exert influence to overcome these problems.

Lack of integration. The effectiveness of emergency planning that has not been integrated into the company's total management system will suffer. Emergency planning should be a key component of any risk management/loss prevention program.

Managers not involved. If managers at different levels through-

out the organization have not engaged in or contributed to the
planning activities early enough in the process the result may be a
number of weaknesses and oversights in the plan.

Expectations too high. Top management is being unrealistic if it
expects immediate results from the planning effort. It takes time and
thoughtful preparation to train personnel and to build strong emergency response capabilities.

Management inflexibility. Top management often does not
realize that emergency planning is an *evolving* process—a company
never completes an emergency plan but rather continually improves
it.

Planning responsibility wrongly placed. If the responsibility for
emergency planning is placed within a single department, the result
can be that other departments ignore the plan.

Too much, too soon. It is to no one's benefit to attempt too much
at once. If the effort is begun cautiously and thoroughly, it can be
speeded up as the process progresses.

Failure to see the big picture. Many departments or managers fail
to see the overall picture of planning and get hung up on the little
details. This can be avoided by establishing definite project goals and
schedules.

Failure to implement the plan. Any of the preceding reasons can
further result in a manager's unwillingness or inability to implement
an emergency plan.

5

Plan Elements: Preliminaries and the Basic Plan

One of the most common questions asked concerning emergency preparedness is, "What should be contained in our plan?" The correct answer is, "Everything that is necessary to meet your objectives." The important thing to remember is that there is more than one acceptable way to meet a certain set of objectives. Format and content of emergency plans vary tremendously among facilities; therefore, choosing the type best suited for your facility is up to you.

As table 5–1 indicates, there are numerous elements that *could* be included in an emergency plan, but those that *should* be in yours depend on what type of plan you have chosen to develop. Chapters 5–7 outline and describe the elements that should be contained in a **comprehensive emergency plan**. If you choose not to develop a comprehensive emergency plan, then refer only to the sections required for the type of plan you choose (use table 5–1 as a guide).

PRELIMINARY ELEMENTS

The plan should be preceded by certain precursory elements, the most self-explanatory of which is a **table of contents**. Discussed in more detail are the four preliminary elements that follow.

Plan Distribution List

A plan distribution list is simply a list of all persons to whom copies of the plan have been assigned. This is necessary so that the plan coordinator will know who should receive updates and revisions to the plan, and it allows management to periodically review the list to ensure that everyone who should have a copy of the plan actually

Table 5–1: Plan Content

Section	Action Guide/ Checklist	Response Plan	Emergency Management Plan	Mutual Aid Plan
Preliminary Elements				
table of contents	O	O	R	R
plan distribution list	O	R		R
record of changes	O	R	R	R
promulgation letter	O	O	O	R
definition of terms	O	R	R	R
Basic Plan Elements				
introduction	O	R	R	R
purpose	O	R	R	R
policy/legal authority	NA	NA	R	R
situations	O	O	R	R
assumptions	O	O	R	R
concept of operations	NA	O	R	R
responsibilities	R	R	R	R
plan maintenance	O	R	R	R
Prevention Procedures				
inspection	NA	NA	R	NA
safety & health reviews	NA	NA	R	NA
Preparedness Procedures				
personnel training	NA	NA	R	R
drills and exercises	NA	NA	R	R
supplies equipment	NA	NA	R	R
protection of records	N	NA	R	NA
mutual aid	NA	NA	R	R
community awareness	NA	NA	R	O
Response Procedures				
detection alert & warning	M	M	R	R
direction & control	M	M	R	R
communications	M	M	R	R
emergency shutdown	M	M	R	R
site evacuation	M	M	R	R
medical treatment	M	M	R	R
government coordination	M	M	R	R
fire procedures	M	M	R	R
bomb threat	M	M	R	R
other hazard procedures	M	M	R	R
Recovery Procedures				
incident investigation	M	NA	R	NA
damage assessment	M	NA	R	NA
cleanup & restoration	M	NA	R	NA
business interruption	M	NA	R	NA
claims procedures	M	NA	R	NA

R = recommended   O = optional   NA = not applicable
M = mandatory, single function procedure

does. The list should contain the holder's name, job title, organization, division (if applicable), telephone number, and date of receipt.

Record of Changes

A record of changes (a sample of which, along with a plan distribution list, is shown in figure 5–1) allows the plan coordinator to record all modifications made to the plan since its original publication. This form simply contains space for a change number, date of change, pages affected by the change, a brief description of the change, and the name of the person(s) who authorized the change. This change form should be distributed along with copies of the pages from the plan that are changed.

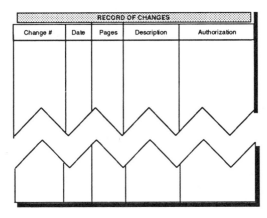

Figure 5–1: Sample plan distribution list and sample record of changes

Promulgation Letter

A promulgation letter, usually a letter or memo from the plant manager or CEO, gives the plan official status and provides authority for the requirements it places on departments within the company. This letter should also contain the date on which the plan becomes effective.

Definition of Terms

A definition of terms is important because clarity is vital to the success of any emergency program. This section should define words, phrases, abbreviations, and acronyms used in the plan, especially terms whose meaning in the context of the emergency plan may differ from everyday usage.

BASIC PLAN ELEMENTS

The term "basic plan" refers to the initial part of the plan, which, in a nutshell, describes how emergency operations will be carried out at the facility. The basic plan answers the questions of who, what, how, and why; it gives purpose, direction, and scope to the rest of the planning process. In addition, this section of the plan often contains other documents pertaining to emergency management such as policy statements and facility and corporate directives. The format of the basic plan should be consistent with the format of other facility documents in order to facilitate employee understanding.

Introduction

The introduction of the emergency plan should be a short paragraph that is a brief abstract of the plan's contents. It should describe the planning process and acknowledge those who contributed to the planning process. It can also identify the intended audience for the plan and how it should be utilized.

Purpose

The purpose section describes the overall goals and objectives of the plan. These should not be "pie-in-the-sky" statements but rather should be based on very concrete and achievable goals. An example of a concrete objective would be: *This plan is intended to provide for a rapid response of the emergency brigade to any emergency situation occurring within the boundaries of the plant—within three minutes of alert.*

Policy and Legal Authority

Following the statement of purpose should be a section concerning policy and legal authority for carrying out the tenets of the plan. This section contains the basis and limits of authority for the plant manager and for other employees in carrying out their duties. It should clearly state the company's policy concerning emergency management. It should also contain a summary of requirements imposed by federal, state, and local statutes. This section should not be a lengthy, legalistic interpretation of the law by the company attorney, but rather a synopsis of legal and company rules in order to make the reader aware of the scope of his or her authority. Specific legal requirements, such as government requirements, can be included in the appropriate subsequent sections of the plan.

Situations

The situations section states all conditions that may be expected to create the need for implementing the emergency plan. Such information is drawn from any hazard analyses performed by the company or by government. This would include not only internal hazards (such as a warehouse explosion) but also external hazards (such as natural hazards or a dam failure). This section should also contain information concerning the local geography and climate, population statistics, details about the local infrastructure, effects of time variables (such as daytime population versus nighttime population), and any other factors that could influence the magnitude of the hazard and the logistics of the response.

Assumptions

Assumptions are those judgments made by the emergency planner when formulating the plan. For example, a planner might assume that a certain amount of inquiries will come into the plant switchboard following an explosion inside the plant, thus creating the need for alternative communications procedures. Assumptions are any judgments made by the planner that cannot be proven or disproven before an emergency and should be included in the plan so that everyone will understand the limitations of the plan.

Concept of Operations

The concept of operations includes a brief description of what is expected to be accomplished if an emergency should occur, and it states how people and organizations will interact during an emergency. This is important because emergency operations can be quite different from normal, day-to-day operations. For example, an assistant plant manager may have a very important decision-making role during normal business activities, yet his involvement may be secondary during an emergency. The concept-of-operations section should also identify and define various levels of alert, or **emergency classification levels**. Each alert level would require a certain type of response; that is, each member of the organization would have to implement certain emergency functions. As the emergency grows in magnitude, the level of response increases.

The nuclear industry effectively uses four alert levels: **unusual event, alert, site area emergency**, and **general emergency**. (Refer to chapter 3.) Plant operators have been trained in what level to declare based on certain plant conditions or events that may occur. When a certain level is declared, all plant staff as well as government emergency agencies follow a planned procedure for that action level. For example, when an unusual event is declared, a specific plant employee knows that he or she must notify certain federal, state, and local officials in accordance with the plan. The benefits of using emergency classification levels is that the response is much more organized and manageable. It also ensures that only those resources necessary to cope with the emergency are employed.

While the Nuclear Regulatory Commission (NRC) defines these action levels for nuclear power plants, there are no designated action levels for most other types of industrial facilities. Therefore, most companies should find some flexibility in identifying and defining action levels appropriate for their own circumstances. For example, one chemical plant utilizes four levels:

An **alert** is declared if there is warning of a possible emergency, such as a hurricane expected to hit within 48 hours.

A **unit emergency** is declared if there is an actual emergency—such as a small fire in one of the plant's three units—but the emergency is not expected to affect the rest of the plant.

A **site emergency** is declared if the emergency affects the entire plant but is not expected to have any off-site consequences.

A **general emergency** is declared if the incident affects both the plant and the surrounding area, such as a major chemical release.

It is important to clearly define under which circumstances the levels would be declared, by whom, and what basic actions would be taken. An example of emergency action levels that could be used in an industrial facility is illustrated in figure 5-2.

ALERT	Internal or external conditions have the potential to threaten the facility.
SITE EMERGENCY	Actual emergency conditions exist at the plant but do not threaten off-site areas
GENERAL EMERGENCY	Actual emergency conditions exist at the plant and have the potential to seriously affect off-site areas

Figure 5–2: Example of emergency classification levels

Responsibilities

Another important section of the basic plan covers the organization and assignment of responsibilities. This section provides the organizational framework for all emergency management activities; departments with responsibilities in any phase of emergency manage-

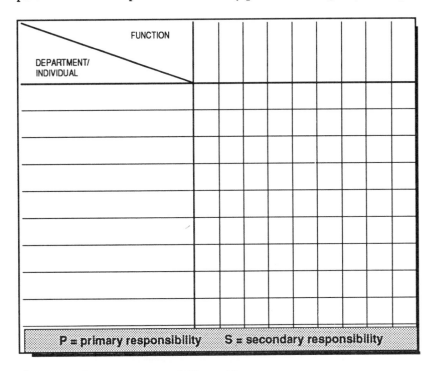

Figure 5–3: Emergency responsibility matrix

ment are identified in this section, along with their clearly delineated functions. Typically, senior managers and departments with major response functions have their roles defined here. This section should also indicate the relationship between various departments in terms of primary and supporting roles. An organizational matrix chart, such as shown in figure 5–3, is useful for graphically displaying these relationships. While very specific responsibilities may be delineated in a later section of the plan, people and departments with major emergency roles should have their overall roles delineated here. Emergency roles and responsibilities should somewhat parallel normal organizational roles. While this is not always possible due to the nature of emergencies and the need for quick decision making, an extreme modification of normal roles is inappropriate. For example, an accounting clerk should not be assigned as liaison to the state environmental affairs department, whereas the plant's environmental affairs manager probably should.

Also part of organizing responsibility is stating a clear chain of command and decision-making authority, as well as identifying alternates for key response functions. At a minimum, one alternate

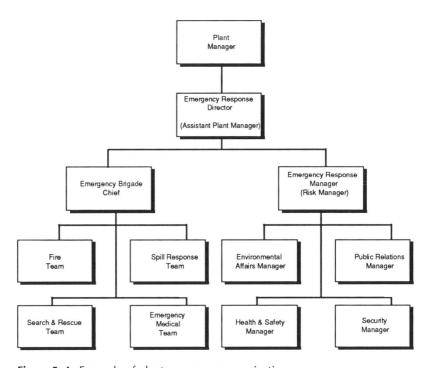

Figure 5–4: Example of plant emergency organization

for each position should be identified; however, two are recommended. This is important because a day hardly goes by where people are not on sick leave, on vacation, out of town for a seminar, or downtown for a corporate meeting. An effective emergency response program cannot be contingent on a select few being at the plant at the right time. When deciding on primary and alternate responsibilities, consider carefully the backshifts. Quite often, the night and weekend shifts do not have the same staffing as does the normal weekday shift; so do not assign major responsibilities to positions that are not filled on a twenty-four-hour-per-day basis. An example of a typical industrial facility's emergency organization is illustrated in figure 5–4.

Plan Maintenance

The basic plan should end with a section about maintenance. This section should outline the procedure for continuing review and improvement of the plan. It should also designate someone to maintain, update, and lead future efforts at revising the plan. Since the emergency plan involves many individuals in many departments and outside agencies, the plan maintenance function requires a person with good coordination skills and a certain amount of authority to maintain the plan.

6

Plan Elements: Prevention and Preparedness

The probability of an emergency occurring can be minimized by implementing effective prevention measures. Since many of the preventive measures designed to control the occurrence of emergencies will also help maximize normal operations, most procedures of the type discussed will be found in other company manuals (e.g., preventive maintenance procedures).

PREVENTION PROCEDURES

Obviously, preventing an emergency from occurring is preferable to reacting to one. Prevention procedure is an important aspect of a comprehensive emergency response plan. Typically, the procedures for **fire prevention, safety and health review**, and **inspection** can be put into this section of the plan. Quite often however, companies prefer to keep these procedures in other existing documents (e.g., a policy and procedures manual). This is fine, but if you are putting together a comprehensive emergency management plan, the prevention elements should at least be incorporated by reference. Responsibility for development of prevention procedures should be assigned only to qualified individuals.

Inspection Procedures

Provisions should be made for inspecting and/or testing critical equipment or components on a regular basis. These procedures should also specify the type (such as visual detailed monitoring or nondestructive testing) and the frequency of inspection or testing.

Examples of the type of equipment or components to be inspected include:

- piping
- pumps
- valves
- fittings
- tanks and containers
- tank supports and foundations
- fire-suppression equipment
- detection-alarm systems

The frequency of inspection should be in accordance with government, industry, and manufacturer standards but at a minimum should comply to the following:

Equipment	Frequency of Inspection
fire-protection equipment	per OSHA suppart L provisions
portable extinguishers	monthly
auto sprinkler system	weekly
yard hydrants and hoses	monthly
tank or reservoir	weekly

Safety and Health Reviews

Procedures should also call for the review of all new processes and equipment for compliance with federal, state, local, or industry standards. Appropriate protective monitoring systems should be installed on high-risk operations. Other questions that should be considered when developing prevention procedures include:

1. Have control measures been implemented to minimize the amount of hazardous materials?
2. Are good housekeeping procedures being used to control fire and health hazards (Ref. OSHA 1910.38b3)?
3. Have security measures been implemented to prevent tampering with critical equipment?
4. Are safety and health programs in use?
5. Do maintenance procedures on safety/protective devices or systems include procedures for ensuring adequate backup protection or procedures to be taken when equipment is down?

6. Are provisions made to protect critical components, safety devices, protective systems, and so on, from deliberate or inadvertent tampering or deactivation?

7. Are contractor/vendor personnel coming on-site advised of or trained in hazards, emergency procedures, and safety procedures, including the use of personal protective equipment?

PREPAREDNESS PROCEDURES

Personnel Training

This section contains guidelines for ensuring that effective training is provided for all emergency response personnel. It covers training procedures to be included in the emergency plan as well as the evaluation of training capabilities of the facility in general.

An effective plan must specify what is to be done, who is to do it, and how it is to be done. The plan will not succeed, however, without trained personnel who have the knowledge and skills necessary to carry out assigned tasks. Providing training that meets the needs of each individual's role within the plan is a key element of successful emergency response.

First of all, the plan should contain procedures for emergency response training that address all types of hazards or emergencies covered by the plan. The emergency response training should be based on these specific hazards and response duties as outlined in the plan as well as on regulations such as OSHA 1910 and the Hazard Communications Act. Regardless of the regulations for training at your facility, training, at a minimum, should cover the following:

- personal protection equipment

- preventive maintenance

- spill response

- first aid

- hazardous materials characteristics

- evacuation procedures

- emergency reporting procedures

- proper fire extinguisher use

- leak warning signs (odor, smoke, sounds, etc.)

Training emergency response personnel (e.g., fire brigade) should provide members with all the knowledge and skills commensurate with their duties. OSHA 1910.120, OSHA 1910.156c and appendix A to subpart L outline many of the requirements for fire brigade members. The people assigned to interior firefighting should be provided with a training session at least quarterly, while other team members should be trained at least annually. Fire brigade officers and instructors should be given a higher level of training.

As with most formal training programs, testing the student's grasp of the required knowledge and skills should be conducted. A fire brigade member's failure to understand his or her emergency response job could be costly in both human and monetary terms. An effective method of emergency training is to incorporate emergency training with normal operating training.

Training should be modified as changes in either hazards or the plan occur. There is a tendency in many facilities to postpone making revisions, but it should become a habitual practice to make these changes as quickly as possible. It is important that all employees with emergency roles be provided periodic retraining and that any new employee be trained within a reasonable time after his or her start date.

Finally, training should be offered to off-site emergency response personnel and on-site fire brigade personnel should take advantage of any off-site training that may be offered by the local fire department. Such joint training not only increases the skill of your fire brigade but also increases the community coordination and the ability for the fire department and the brigade to work together.

The plan should also contain provisions for administering the training program. An individual should be designated to administer the emergency training program and to develop new programs. A prime responsibility of this person should be to continually review the adequacy of the training one method of which is through the use of drills and exercises. This person should also be capable of determining the minimum training levels for all emergency response positions. Record-keeping procedures for the training program should be outlined in the plan or in a related operating procedure, and should specify documentation requirements, including the student's name, type of training, date of training, and results of testing.

Drills and Exercises

While drills and exercises are often used as effective training tools their primary importance lies in their effectiveness as a means of testing emergency response capabilities. They provide virtually the only means, short of an actual incident, of measuring the state of readiness and of testing the effectiveness of an emergency response plan. The plan should contain procedures on the types and frequencies of exercises and on organizing, conducting, and evaluating them. (See chapter 13 for extensive information on developing an effective drill and exercise program.) The plan should, at least, delineate the following:

1. Responsibility for developing, scheduling, and conducting drills/exercises.
2. Provisions for including all levels of management in the exercise program.
3. Provisions for involving off-site personnel/agencies in the drills.
4. Provisions for correcting defects in the plan that are detected by the drills/exercises.
5. Provisions for conducting an annual full-scale exercise.
6. Provisions for drills on the following key elements:
   - communications
   - fire control
   - medical first-aid response
   - spill control
   - emergency operations center
   - monitoring
   - cleanup
   - evacuation

Supplies and Equipment

The emergency plan should contain procedures for controlling and maintaining special supplies and equipment for emergency response. In addition, it should contain an inventory list as well as procedures to ensure that all equipment inventories are kept current. Equipment maintenance procedures based on standards of the manufacturer should be established. The plan should provide for emergency personnel to have easy access to equipment; however, there should be enough security to ensure that access by unauthorized personnel is minimized. The personal protection equipment and other emergency response supplies should be kept separate from normal operating supplies. Of utmost importance, there should be lists of outside sources of emergency supplies and equipment mentioned in the plan (or incorporated by reference).

Protection of Records

Many plans neglect to include procedures for records protection, even though protecting vital statistics and other recorded information from loss is essential. Even a small fire or minor emergency could have disastrous effects on a facility if business records are lost. Any comprehensive emergency plan should identify all records that are considered vital. Examples of important records include employee records, customer records, and financial information. If any of these records are protected by special measures (e.g., safes or Halon systems), their locations should be identified. For those records not protected by such measures, duplicate records and/or off-site storage should be considered. The emergency planner should check with individual departments (e.g., data processing, personnel, purchasing, engineering) to determine if they have already developed their own records-protection procedures that should be included in the plan.

Mutual Aid

Due to the variety and complexity of industrial hazards, it is often desirable to have several facilities band together in order to maximize protection while minimizing costs. Mutual aid refers to agreements to share emergency resources such as equipment, information, personnel, and possible financial assistance during emergencies.

If any mutual-aid agreements (written or verbal) exist, they should be attached to the plan or incorporated by reference. At very least, the plan should contain procedures for implementing the agreement during the emergency. The plan or the agreement should clearly define the level and type of support to be given by each party. A listing of capabilities and resources for all facilities party to the agreement should be included in the plan. The procedures should also outline under what conditions assistance would be rendered and the methods for coordinating communications and response activities between or among facilities.

Mutual-aid agreements should be reviewed by attorneys so that all liabilities and methods/rates of compensation are known and approved of in advance. It is important to remember that cost and other conditions should not be so prohibitive that member facilities would fail to utilize the resources available to it under this program.

7

Plan Elements: Response and Recovery

RESPONSE PROCEDURES

Detection Alert, and Warning

Detection systems are automatic devices for monitoring chemical leaks or early warning devices, such as smoke detectors, which are intended toward emergency situations. They are sometimes referred to as alarm systems, although some alarm systems are not detection systems (e.g., fire-alarm manual "pull stations"). The use of detection equipment varies from plant to plant, but normally each plant has a means of automatically sensing an abnormal condition. Detection systems are normally designed around a single type of emergency condition or hazardous substance; however, the use of multiple types of sensors can be integrated into a single system as is commonly done with fire/security alarm systems.

It is important that the planner consider detection devices within the emergency plan. Early warnings can evoke early responses, sometimes before a situation has the chance to develop fully and get out of control. The plan should provide for the use of any detection system that is applicable to the facility's potential hazards. Among the more common early warning devices available are the following:

- smoke detectors
- heat detectors
- remote single-substance monitors
- leak detectors
- process-control alarms

These detection devices should be monitored continuously, and the plan should contain provisions for regularly testing, inspecting, maintaining, and calibrating these devices. Procedures should also be developed for interpreting, reporting, and responding to data from detectors or process controls, and training should be provided regarding the interpretation and response of data from detection devices.

After detecting the presence of an emergency, the first step is communicating information about the emergency to response personnel and warning all personnel in the general area that may be endangered by the emergency situation. The methods used for alerting/warning may vary widely from facility to facility; they depend on many factors, including the types of communications available and management's basic response policy.

When discussing emergency notification, the plan should identify the contact points that employees should alert if they discover an emergency. Many facilities designate a central point, such as a switchboard or a security guard, to receive reports of an emergency. Procedures should also be developed for then alerting appropriate response personnel via telephone listings, radio beepers, and/or plant intercom/page systems. Procedures should be established for issuing a general warning to other plant personnel via plant-wide alarm systems, coded alarm systems, plant intercom/page systems, and/or telephone. In addition to a primary method, an alternate method should also be designated. When using telephones, be sure that procedures clearly state that emergency messages be given priority. Provisions should also be developed for alerting the following in the event of an emergency:

- remote and/or noisy areas
- plant visitors
- off-site personnel/departments
- off-shift personnel

Alerting systems must be maintained and tested on a regular basis, at least every two months according to OSHA.

Emergency notification procedures should outline the content of off-site notification messages. These messages should be designed to answer the following seven questions:

- What happened?
- To whom did it happen?
- How did it happen?
- What help is needed?

- Where did it happen?
- When did it happen?
- To what extent?

A flow chart of the public alerting process as depicted by the EPA in its "Review of Emergency Systems" report to Congress is shown in figure 7–1.

Direction and Control

Control methods must be established in order to quickly organize a complex assortment of personnel and resources into an effective response. The usual, but not exclusive means of performing the direction-and-control function is to establish a control point, or emergency operations center (EOC). The EOC can best be described by comparing it to a military command post, complete with staff, communications and, protection. The guidelines outlined in this section are intended to assist the industrial emergency planner in developing procedures for the proper direction and control of emergency response activities.

Some key considerations in developing an EOC are as follows.

1. A set location for the control point should be designated.
2. Alternate locations should be specified.
3. Primary and secondary locations for the control center should be convenient.
4. Locations must provide an adequate measure of protection from hazards (e.g., being separate from a building with high fire risk).

The EOC must be adequately equipped with the following:

- communications equipment, including secondary means
- warning devices, such as alarms
- protective equipment for staff, including first-aid kit
- technical information on hazards, maps, etc.
- up-to-date documents, including telephone listings, letters of agreement with off-site agencies, etc.
- emergency power source for lighting and other utilities
- administrative office supplies
- personal necessities (e.g., food, water, and toilet and sleeping facilities) adequate for expected length of stay

Standard EOC operating procedures should provide for:

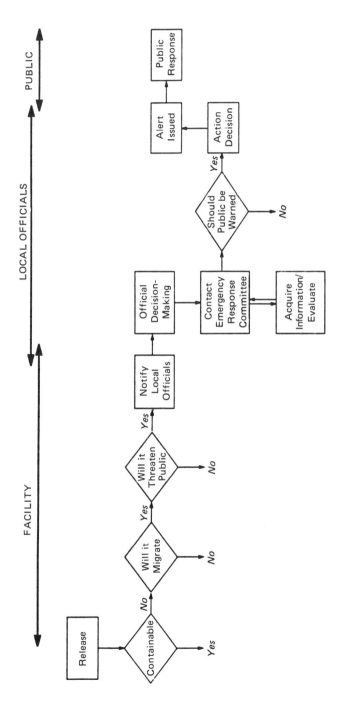

Figure 7–1: Public alert process

- activation of center, including notification of staff
- on-site communications
- off-site communications
- use of equipment/technical support
- press information/public information
- accident assessment capabilities

A specially trained staff should be designated, with individuals assigned in the following functional areas:

- policy
- situation analysis
- operations
- communications
- hazards monitoring
- meteorological monitoring
- technical support
- clerical and administrative services
- press/public information

Security and access control provisions should also be developed to prevent unauthorized individuals from interfering with EOC staff. Procedures should be established and individuals assigned responsibility for maintaining EOC equipment in a state of readiness.

Communications

Effective communications have been proven time and again to be the key to effective emergency response. The ability of the various emergency organizations to adequately respond, coordinate, report, and make requests depends on effective communication with other groups. Emergency notification differs from communication in that the former refers to initial warning messages and the latter refers to systems necessary to conduct ongoing emergency operations. This section provides guidance for the establishment of effective emergency communications procedures. While the material in this section deals with both equipment and procedures, technical advice and assistance in the selection of appropriate communications equipment should be sought.

Communications systems such as two-way radios, intercompaging, telephones, or even runners, should be available for emergency communications. Backup and redundant systems should also be developed. Provisions should be established for communications between the following groups:

- EOC and response teams
- EOC and all off-site agencies
- on-site response teams and off-site response teams
- EOC and support personnel, including press/public relations and technical support (e.g., CHEMTREC)

Operating procedures should also include provisions for:

- prioritizing messages
- documenting messages
- tracking messages
- utilizing a radio protocol

As with all emergency equipment, procedures for maintaining and regularly testing communications equipment should be written.

Emergency Shutdown Procedures

In making provisions for an emergency shutdown of a facility's processes and equipment, the planner should realize that there is a two-fold objective: The first is to ensure the proper and thorough valve-shutting and power-cutting to all systems that are nonessential to the emergency response process and/or could escalate emergency conditions. The second is to ensure that no essential controls are shut down. The closing of some valves or the shutting off of power to a critical piece of equipment could hinder response actions and create other hazards as well.

Specific individuals should be designated and assigned responsibility for emergency shutdown activities. It is very useful to have checklists available for each individual operation such as equipment or plant area to be shut down. These checklists should be readily available to both operating and response personnel. Diagrams and maps showing where critical components are located should be attached to the procedures and checklists. Critical components, valves, and controls should be clearly indicated by color coding or similar markings.

Fire and Spill Procedures

Procedures for the emergency response team to respond to fires and spills must be more specific than other emergency procedures. Up to this point in the review and writing of the plan, the emphasis has been to develop procedures flexible enough to use in responding to a wide range of circumstances. Fire and spill response procedures for the response team are different in this regard since specific emergency conditions must be considered. The plan must contain provisions for the emergency response team to deal with:

- fires
- chemical spills
- explosions
- atmospheric releases
- other hazards

Each plant should have a fire brigade or response team specially trained and equipped to deal with specific hazards. Each team member should have specific tasks assigned and should be fully trained to perform those tasks. While membership might be voluntary, every member's attendance at training, drills, and meetings must be mandatory. Procedures should state the minimum manpower requirements, based on tasks to be performed, for each shift's emergency team (up-to-date rosters should be available). A responsible individual should be appointed as a supervisor for each team and on each shift.

Standard operating procedures should include the following:

- clear definition of chain of command, method of responding, and breakdown of tasks
- coordination procedures with other facility personnel
- coordination procedures with off-site organizations
- determination of whether chemical substances are involved
- proper use of personal protection equipment
- measurement and marking of "hot" zone and access control
- accounting for response personnel when entering or leaving a "hot" zone
- training, drills, and related meetings, including participation of off-site organizations

If hazardous materials are produced, stored, or used at the site, spill cleanup and disposal procedures should include:

- determining what debris should be removed
- identifying disposal sites and transportation
- determining proper containers for storage and transport

Facility Evacuation

In an emergency situation, it may be necessary to evacuate some or all areas of a facility in order to protect personnel. Planning must consider primary and alternate routes of evacuation, methods of informing employees, and account of personnel. This section is intended to address facility evacuations only and not community evacuations.

The plan should include provisions for the emergency evacuation of personnel from all plant areas. A prearranged alarm or signal should be designated to notify employees of the need to evacuate. A responsible individual should be assigned authority to order a full evacuation. Conversely, the "all clear" order to return should be given only by an authorized individual.

Designate an evacuation coordinator and divide the plant into logical evacuation zones. Among the evacuation coordinator's duties should be to guide others to evacuation/exit routes, to check the area for strangers, to turn off noncritical equipment, and to close windows and doors.

Assembly areas where evacuated employees are to report should be designated and specified in the plan. These areas should be located at safe distances from plant facilities that might become engulfed by the hazard agent. Evacuation maps and/or instructions should be conspicuously posted throughout the plant.

Procedures should be developed for each area to account for and report on the presence of personnel. This information should be forwarded to the EOC for compilation and evaluation. If an employee does not report to his or her designated evacuation area, he or she might have reported to another area, and the EOC, as a center of information, would be in a good position to determine this. The plant evacuation process should also provide for an accounting of visitors and special procedures for the handicapped.

While rarely necessary, provisions should be made for the transportation and temporary shelter of employees. The need for this provision will vary greatly from facility to facility, and most will rely on employees to transport themselves out of the area (presumably to go home) if necessary, but certain facilities might find it necessary to seek alternative solutions.

Finally, as with any critical procedure, the plan should mandate that all employees be provided with evacuation training and participate in at least one evacuation drill per year.

Security Considerations

The security function is one of the most important support functions for emergency response. While not directly related to specific emergency mitigation tasks, failure to control access to the plant, control traffic flow, and protect vital records and equipment from tampering can result in confusion and less-than-adequate response.

Prior to developing the emergency security procedures, the following factors should be considered by the planner:

1. Does the facility have in-house security personnel? Contract security personnel? No security personnel?
2. Does the plant have physical access control barriers, such as fences and alarms?
3. Are the facility's vital emergency resources adequately secured?

Provisions should be made to control access to the facility during an emergency, especially regarding the following key areas:

- EOC/control point
- media center
- emergency supply storage areas
- strategic areas, such as research and development

Provisions should be made for controlling traffic in and around the facility, and this function should be clearly coordinated with off-site public safety officials. Procedures for controlling pilferage during and after an emergency is another consideration to be developed in advance.

Public Relations and Emergency Information

The emergency plan should specify how public relations should be implemented before, during, and after an emergency. While academic research fails to prove just how important community education programs are for ensuring effective public response during emergencies, it is clear that many segments of the population appreciate having emergency information prior to an emergency. Widely used in industry are public information documents that contain the following:

- hazards information

- process/facility description
- consumer products produced
- safety record
- emergency plan summary
- environmental protection policy
- company contributions to community
- annual report/financial data

This type of information should not only be distributed to the general public through printed matter such as brochures, but should also be distributed to the public through employees, managers, news media, and civic/community groups.

The plan should identify a specially trained individual who will be responsible for the pre-emergency information program as well as for public information during an emergency. This public relations person should be among the first to be called during an emergency, and the plan should contain provisions requiring the EOC to keep public-relations (PR) personnel up to date. The procedures should include the establishment of a press area and should contain the general policy on information to be released, administrative procedures for logging press/public inquiries, and other PR functions.

Coordination between Facility and Off-site Agencies

Coordination between the facility and off-site organizations is essential because no plant can be considered fully self reliant, and because under severe circumstances, plant emergencies can affect the general public. The plan should contain a listing of outside agencies/organizations along with their emergency response capabilities. The coordination methods should be clearly specified in the plan, and written agreements, if available, should be incorporated by reference into the plan. Groups such as fire and police departments should become familiarized with the facility. In addition, the plan should contain a telephone listing of contacts, along with information to be given to each organization at the time of emergency.

Written operating procedures should outline provisions for:

- communications
- compensation
- type of support
- liability limits
- contacts

- circumstances requiring assistance
- technical/hazards information

RECOVERY PROCEDURES

This section provides planning guidance in developing postdisaster recovery procedures designed to minimize the effects of a disaster on business operations. **Incident investigation, damage assessment, cleanup and restoration, business interruption,** and **claims procedures** are all subjects to be included in recovery procedures.

Due to the potentially devastating effect of a severe emergency on management structure, the plan should contain provisions for a clear line of succession for all top management and key personnel. Since a plant may be unihabitable following an emergency, the plan should specify an alternate location for management operations. The plan should provide for the establishment of a recovery control team to coordinate recovery operations.

The plan should specify procedures for preserving the accident scene (or impacted area) for accident investigation by on-site or off-site personnel. It should also contain procedures, consistent with insurance requirements, for documenting all compensable losses. Recovery preparations should include:

1. Assignment of personnel to supervise cleanup and repair.

2. Notification procedures to inform personnel not to report to work as scheduled.

3. Damage assessment procedures.

4. A prioritized list of repairs or replacement of critical equipment.

5. Special procedures to expedite issuance of work orders, purchase orders, etc.

6. A list of available cleanup equipment, including quantities and locations.

7. A designated area for the temporary storage of damaged equipment and materials until released by insurance investigation.

8. Special accounting procedures to ensure accurate loss figures.

9. Provisions for accident investigation and response critique to be used to upgrade plan.

10. Procedures for measuring levels of contamination and safe levels for reentry.

8

Evaluating, Reviewing and Maintaining the Plan

There is an adage that says an emergency plan is outdated the day after it is completed. There is a lot of truth to that since businesses are constantly changing—processes change, staffing levels change, community capabilities change, and so on. While the basic tenets of an emergency plan probably remain valid for some time, the plan should be continually reviewed and maintained.

It is not enough to review a plan by reading through it looking for minor errors and typos. The plan should be thoroughly evaluated through a series of drills and exercises (discussed in chapters 13–15). This will point out deficiencies in the plans and procedures, as well as in training and capabilities. The plan should also be reviewed following actual emergencies to determine how well it worked under actual operating conditions. Changes should be made to all aspects determined to be ineffectual.

All emergency plans become outdated because of social, economic, and environmental changes. Keeping the plan current is a difficult task, but can be achieved by scheduling reviews regularly. The plan should indicate who is responsible for keeping it up-to-date. Outdated information should be replaced, and the results of appraisal exercises should be incorporated into the plan.

The following procedures will aid in keeping the plan up to date:

1. Establish a regular review cycle. This review should occur every 6–12 months.

2. Assign one individual the responsibility to maintain the plan.

3. Include a "record of changes" in the plan so that all changes after its initial issuance can be recorded.

4. Make sure that all holders of the plan know to whom to report changes.

5. Keep plans in a three-ring binder to allow for easy changes.

6. Check all information likely to change with time, such as telephone numbers, names, equipment locations and availability, addresses, etc. In addition, ask departments and agencies to review sections of the plan defining their responsibilities and actions (**Note:** The plan should use titles, as opposed to the names of individuals with emergency response functions. This will eliminate the need for frequent changes to the plan.)

7. Distribute changes to all holders of the plan. Changes should be consecutively numbered for ease of tracking. Be specific (e.g., "Replace page ___ with the attached new page ___...." or "Cross out _____ on page ___ and write in the following... [e.g., new phone number, name location, etc.]."). Any key change (e.g., new emergency phone number, change in equipment availability, etc.) should be distributed as soon as it occurs. Do not wait for the regular review period to notify plan holders. Request an "acknowledgment of receipt" from persons who receive the change.

8. Make sure that recommendations included in incident reports and exercise evaluations are incorporated into the plan in a timely manner.

9. Make sure that emergency training courses are modified to comply with any modifications in the plan.

9

Integrating Facility Plans with Community Plans

Local governments are responsible for developing plans to protect the public in event of emergency. This responsibility stems from a variety of state and local laws that require emergency planning. Some laws are very specific and detail what type of planning must take place, while others may be very general, leaving the extent of planning up to local leaders. Most states have civil defense and emergency preparedness laws that mandate the development of community emergency plans. You can check with your town clerk, city attorney's office, or civil defense director if you would like specific citations for laws that affect your community. Title III of the Superfund Amendments and Reauthorization Act (SARA) also mandates the development of local plans to deal with hazardous materials emergencies. Suffice it to say, most community officials feel it is the community's responsibility to plan for emergencies in order to provide for the general welfare of the public.

Until recently, authority for "hazmat" (hazardous materials) planning has been a bit vague. A study published by Quarantelli* in 1984, *Chemical Disaster Preparedness at the Local Community Level* made some interesting observations. Some of the disclosed paradoxes in the local-level planning for chemical emergencies included the following:

1. Chemical facilities that engage in the most planning are considered by the community as the ones that do not need to plan as much (e.g., large modern plants versus smaller, poor facilities).

*Quarantelli 1984 (E. L. Quarantelli, Ohio State University, published in the Journal of Hazardous Materials, Vol. 8 No. 3)

2. Chemical facilities generally try to contain on-site emergencies and often fail to notify the community. If a plant is unable to contain an emergency, the community feels it is at greater risk because of the lost warning time.

3. Unlike the case for natural disasters, there is no one community organization responsible for both planning for and responding to chemical disasters.

4. If an agency in a community takes responsibility for hazardous materials emergency planning, other agencies tend to slack off.

This situation may have been rectified by the establishment of State Emergency Response Commissions (SERCs) and Local Emergency Planning Committee (LEPC) under Title III of SARA. Only time will tell if this will remedy Quarantelli's concerns.

It is vitally important that community officials be involved in the company's emergency planning process and that the company be involved in local planning. The facility's plan is more likely to include the correct procedures for notifying and coordinating with government agencies during emergencies if these local agencies are involved. Community involvement during the planning stage will also enable local officials to assess their own capabilities to respond to an emergency at the local facility. Finally, these officials will have a better understanding of the facility's processes and will gain additional confidence in the facility's ability to control these operations. An illustration of how industrial facilities and communities integrate before, during and after an emergency is depicted in figure 9–1.

Regardless of the past level of community activity in your planning process, remember to at least try to include officials in current and future planning efforts. Never "spring" a new plan on community officials, since they are likely to resent it.

Some suggestions for involving local officials in your planning process are as follows:

1. Call the various public-safety agency heads and ask them for a meeting.

2. Discuss your intentions to develop a facility emergency plan and preparedness program.

3. Ask for their advice and assistance during this process.

4. Ask that they assign specific staff personnel to help in the event of a department head's absence.

5. Hold periodic (but not too frequent) meetings during the assessment and development phases.

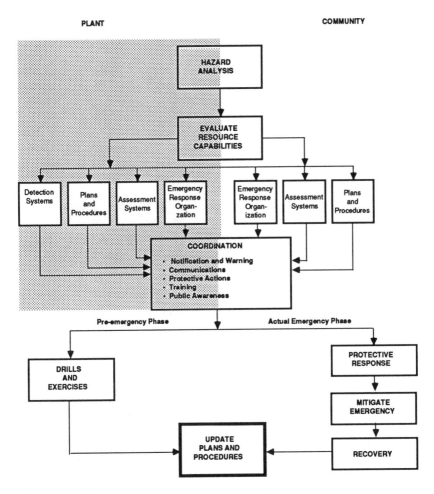

Figure 9-1: Plant/community emergency integration

6. Seek their advice, and follow through with their suggestions. If you are unable to implement their suggestions, explain to them why not.

7. Ask for their concurrence at the completion of the planning project.

8. Have your senior management personally thank them for their assistance, and send letters of appreciation to their superiors.

9. If they in turn ask for your assistance on a special project, try to help them.

10. Involve these officials and their departments in annual training and exercise programs.

ASSISTING WITH COMMUNITY
EMERGENCY PLANNING

The National Response Team's NRT-1 recommends various elements of local emergency planning. The remainder of this chapter
contains several excerpts from the NRT-1, many of which are
followed by comments and suggestions to further enable you and
your facility to be of help in writing various sections of the local plan,
and/or to ensure that the community has adequate capabilities for
carrying out its provisions.

1.0 INCIDENT INFORMATION SUMMARY
The [local emergency] Plan should contain detailed description
of the essential information that is to be developed and recorded
by the local response system in an actual incident, e.g., date and
time, location, type of release and material released.

2.0 PROMULGATION DOCUMENT
The Plan should contain: A document signed by the chairper-
son of the LEPC, promulgating the plan for the district; and
documents signed by the chief executives of all local jurisdictions within the district; letters from covered facilities endors-
ing the plan.

3.0 LEGAL AUTHORITY AND RESPONSIBILITY FOR RESPONSE
The Plan should: Discuss legal authorities of the jurisdictions
whose emergency response roles are described in the plan,
including authorities of the emergency planning district and
the local jurisdictions within the district; List all other authorities the LEPC regards as essential for response within the
district, including state and federal authorities.

While it can be assumed that the majority of communities have
the assistance of legal council in researching and writing this section
of the plan, it is not uncommon for only minimal input to be received
from lawyers. Also, without a fair amount of research, many obscure
laws, regulations, or executive orders are overlooked. For example,
many plans contain references to major state and local civil defense
laws, but rarely do they cite executive orders that give local officials
a strong degree of power during emergencies.

Even when all applicable laws regulations, and executive orders
are cited, rarely are they explained in depth. This fact is supported by
research of previous emergencies in which many local officials really
did not understand the scope of their authority. In some cases,
officials did not exercise their power to the extent possible; in others,

they overstepped their authority. A question often asked is, "Do I have the authority to order people out of their home during an evacuation?" It is unconstitutional to order someone out of his or her home during an evacuation but if found outside his or her home, that person may be evacuated out of the area if an official state of emergency has been declared. Despite this fact, several communities have passed local ordinances making it a crime to stay in one's own home if an evacuation has been ordered.

It should be obvious that the careful preparation of this section of the plan is very important and should not be put off until disaster strikes. Your company could help the community by doing some of this legal research, especially if your community can ill afford to hire its own lawyer for this purpose.

4.0 TABLE OF CONTENTS

The Plan should: List all elements of the plan, provide tabs for each and provide a cross reference for all of the nine required elements of SARA, Section 303(c). Plans that are prepared in the context of requirements of CPG 1-8, should contain an index to the location of both NRT-1 and Title III elements. If it is not done by the LEPC, it should be done by the SERC.

5.0 ABBREVIATIONS AND DEFINITIONS

The Plan should explain all abbreviations and define all essential terms.

6.0 PLANNING FACTORS

Assumptions: Assumptions are the advance judgments concerning what would happen in the case of an accidental spill or release. The Plan should list all of the assumptions about conditions that might develop in the district in the event of accidents from any of the covered facilities or along any of the transportation routes.

Planning Factors: The planning factors consist of all the local conditions that make an emergency plan necessary. Identify and describe the facilities in the district that possess extremely hazardous substances and the transportation routes along which such substances may move within the district. Identify additional facilities in the plan because of their vulnerability to releases from facilities with extremely hazardous substances. Identify other facilities included in the plan by virtue of their potential for releases of hazardous substances; Include methods for determining the occurrence of a release, and the area of population likely to be affected by such release.

Include the major findings from the hazard analysis, (date of analysis should be provided) including: major characteristics of covered facilities/transportation routes impacting on the types and levels of hazards posed, including the types, identities, characteristics and quantities of hazardous materials related to facilities and transportation routes; Potential release situations with possible consequences beyond the boundaries of facilities, or adjacent to transportation routes.

[Include] maps showing locations of facilities, transportation routes, and special features of district, including vulnerable areas.

The Plan should include: Geographical features of the district, including sensitive environmental areas, land use patterns, water supplies, and public transportation networks; Major demographic features of the district, including those features that most impact on emergency response, e.g., population density, particularly sensitive institutions, and numbers of persons with particular disabilities such as deafness; the district's climate and weather, as they affect airborne distribution of chemicals; and critical time variables impacting on emergencies, e.g., time of day and month of year in which they occur.

The development of this portion of the plan, which is based principally on the outcome of the hazard analysis study, is a key area where you can contribute. With the variety of analysis methodologies available, your involvement in this process will help ensure that the community's plan is based on realistic assumptions about your plant.

7.0 CONCEPT OF OPERATIONS

The Plan shall designate a community emergency coordinator and facility emergency coordinators, who shall make determinations necessary to implement the plan. The Plan should identify by name and organizational affiliation the community emergency coordinator and each of the facility emergency coordinators; explain the relationship between these coordinators, their organizations, and the other local governmental response authorities within the district (e.g., the county emergency management authority); describe the relationship between this plan and other response plans within the district which deal in whole or in part with hazardous materials emergency response e.g., the county Emergency Operations Plan and plans developed by fire departments under OSHA

Regulation CFR 29 Part 1910-120; list all the facility emergency plans that apply to hazardous materials emergency response, including all plans developed under OSHA Regulation on Hazardous Waste Operations and Emergency Response (CFR 29 CFR Part 1910.120); describe the functions and responsibilities of all the local response organizations within the district, including organizations such as the Red Cross; describe the relationship between local and state emergency response authorities; describe the way in which these plans are integrated with local response plans; list mutual aid agreements or other arrangements for sharing data and response resources; describe conditions under which the LEPC will coordinate its response with other districts and the means or sequence of activities to be followed by districts in interacting with other districts in emergency conditions; describe the relationship between plans of the district and related state plans; describe the relationship between emergency response plans and activities in the district and response plans and activities by federal agencies, including all plans and responses outlined in the National Contingency Plan.

This section clearly suggests that you must provide certain information to the LEPC, namely giving the name of your facility's emergency coordinator and other facility plan information. It is important that you involve yourself with the writing of this section because it involves the integration of your facility's on-site response forces with the off-site, community forces.

10

Improving Community Planning and Response Capabilities

As does the previous chapter, this chapter contains some excerpts from NRT-1, the National Response Team's guide for developing an emergency response plan. Following most excerpts, ideas and suggestions for improving community plans and response capabilities are presented. It is in your company's best interests to help the community in this way for public-relations purposes as well as for other common-sense reasons. Suppose an emergency occurs at your facility and the emergency response is ineffective; if word leaks out that your facility has done little with community preparedness people, a public backlash might result, hampering your future business plans.

These recommendations are suggested to, in the long run, avoid expensive, mandatory contributions to the community. Your voluntary assistance now could help to avoid what has happened to the nuclear industry. Following the Three Mile Island Nuclear Power Plant accident in 1979, the federal government passed stringent planning regulations that require both plants and communities to have extensive preparedness programs in place before operating the plants. Since few communities had any "real-life" motivation to do more planning than they had already done, nuclear-plant owners had to pick up the slack and pay millions to improve local capabilities just to operate their plants. Other industries are not yet faced with this dilemma, but a few well-spent dollars now will not only help your community be prepared for all types of hazards, it may help to avoid large, mandatory expenditure in the future.

Many industrial emergency planners resent this recommendation to invest in what they perceive to be a basic government function that is already supported through company taxes. But the fact of the matter is that most elected officials have not invested in good emergency programs, and facility-specific planning is very expensive. Most government planners know what has occurred in the nuclear industry, and some perceive SARA's Title III and other recent laws as opportunities to obtain funds from industry. Like it or not, there is precedence for industry to pay for building local emergency response capabilities, and if your facility does not make a voluntary investment now, it may be forced to later.

Sections 10 and 11 call for procedures to be developed to ensure proper notification of all officials in case of emergency.

10.0 NOTIFICATION PROCEDURES

The Plan shall include procedures providing reliable, effective, and timely notification by the facility emergency coordinators and the community emergency coordinator to persons designated in the emergency plan, and to the public, that a release has occurred.

The Plan should include an Emergency Assistance Telephone Roster containing an accurate and up-to-date list of telephone numbers for the: Technical and response personnel; Community emergency coordinator, and all facility emergency coordinators; CHEMTREC; National Response Center; Participating agencies; Public and private sector support groups; Community emergency coordinators in neighboring emergency planning districts; and the Points of contact for all major carriers on transportation routes within the district.

11.0 INITIAL NOTIFICATION OF RESPONSE AGENCIES

The plan should include procedures for notifying the appropriate 24-hour hotline first, and which should be located in a prominent place in the plan and describe methods or means to be used by facility emergency coordinators (FECs) within the district to notify community emergency coordinators (CECs) of any potentially affected districts, and SERCs of any potentially affected states, and any other persons to whom the facility is to give notification of any release, in compliance with Section 304 of Title III;

It should describe methods by which the CECs and local response organizations will be notified of releases from transportation accidents, following notification through 911 systems or specified alternative means and describe methods by which the CEC, or his designated agent, will ensure that contents of notification match the requirements of Section 304, including the regulations contained in 40 CFR Part 355 (Notification Requirements, Final Rule);

The plan should list procedures by which the CEC will assure that both the immediate and follow-up notifications from facility operators are made within the time frames specified by Notification of Final Rule in 40 CFR Part 355; identify the person or office responsible for receiving the notification for the community emergency coordinator or his designated agent and list the telephone number; list the 24-hour emergency hotline number(s) of all the emergency response organization(s) within the district; list all local organizations to be notified and the order of their notification, and list names and telephone numbers of primary and alternate points of contact; list all local institutions to be notified of the occurrence of a release and the order of their notification; list all state organizations to be notified; and list all federal response organizations to be notified, including the National Response Center.

In most industries, helping communities develop proper notification should not require too much involvement other than to provide information concerning the facility's key emergency communicator and coordinator. If your plant has quickly developing hazards and the community has a poor notification system, then an investment in a "prompt notification system" might be in order. Nuclear facilities must demonstrate that all key local and state officials be notified within 15 minutes of an accident, and many have purchased or leased pager systems for local officials, in order to facilitate rapid notification. Recently, there has also been a trend towards the use of cellular telephones for key officials. The main issue is to determine, based on the your facility's hazards and risks, how much time it should take to notify local officials of an emergency at your plant and then develop systems accordingly.

Another NRT recommendation for community plans concerns direction and control of resources during an emergency.

12.0 DIRECTION AND CONTROL

The Plan shall: include methods and procedures to be followed by facility owners and operators and local emergency and medical personnel to respond to any release of such substances. It should identify the organization within the district responsible for providing direction and control to the overall emergency response system described in the Concept of Operations; identify persons or offices within each response organization who provide direction and control to each of the organizations; identify persons or offices providing direction and control within each of the emergency response function; describe persons or offices responsible for the performance of incident command functions and the way in which the incident commander concept is used in hazardous materials incidents; describe the chain of command for the total response system, for each of the major response functions and for the organization controlled by the incident commander; identify persons responsible for the activation and operations of the emergency operations center, the on-scene command post, and the methods by which they will coordinate their activities.

The plan should list three levels of incident severity and associated response levels and identify the conditions at each level.

Defining the concept of operations is a key part of the emergency planning process and one in which you should be very involved. No equipment purchases can help improve this process; only sound thinking can. One technique that is useful for developing this concept is flow-charting. Flow charts make it easier to spot holes or deficiencies in the logic. There are several flow-charting software packages on the market, and even a simple computer-aided drawing program can meet this objective. If the community does not have this capability, then you could donate such a system or have the flow-charting done at your facility following each meeting for review with local officials. A typical community incident-command organization is depicted in figure 10–1.

The NRT recommends that an effective community communications system be established:

13.0 COMMUNICATION AMONG RESPONDERS
The Plan should describe all the methods by which identified responders will exchange information, communicate with each other during a response—including the communications networks and common frequencies to be used; [At a minimum, these methods should be described for each function. Both communications among local response units and between these units and facilities where incidents occur should be described]. It should describe the methods, including computers with online data bases by which emergency responders can receive information on chemical and related response measures.

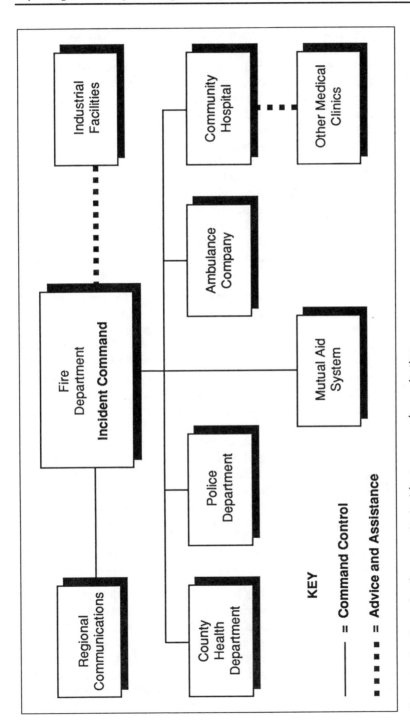

Figure 10–1: Example of community incident-command organization

Recent studies have shown that a key element of an effective emergency response is communications, especially when an evacuation is necessary. A key deficiency that usually is identified in most drills and exercises is communications. With this in mind, you should try to determine the effectiveness of local communications systems and what could be done to improve it. Quite often, communications problems are caused by procedural deficiencies, not by equipment problems, so therefore it is better to review all communications procedures and protocol before investing in new equipment. If there are equipment deficiencies, there may be ways for the community to obtain 50% matching federal grants to purchase new hardware. Of course the local community needs to put up the other 50%, and they may come to you for the match, but in the long run, it may be the best expenditure you could make.

Section 14 of NRT-1 discusses the need for warning and notification systems:

14.0 WARNING SYSTEMS AND EMERGENCY PUBLIC NOTIFICATION
The Plan should describe the methods by which the community emergency coordinator and all responsible authorities within the district will notify the public of a release from any facility or along any transportation route, including sirens and other signals, other alerting methods, use of the broadcast media and the Emergency Broadcast System. This should include a description of the sirens and other signals to be employed, their meaning, their methods of coordination, and their geographical coverage; other methods, such as door-to-door alerting that may be employed to reach segments of the population that may not be reached by sirens or other signals; time frames within which notification to the public can be accomplished; describe the roles and organizational position of the public information officer during emergencies; describe methods for the coordination of the provision of information to the public during a response; and describe any activities by covered facilities in the notification of the public, not covered above.

Up until this point, it has been recommended that you invest in the community's response capabilities by assisting in the purchase of some new equipment as necessary, but this does not include purchasing a community siren system since large siren systems may not be the best idea for your community. Many believe that community siren systems are the single best method of warning the public during an emergency. However, while sirens are very effective, they are not the only warning method. In most of the largest evacuations that have occurred in the United States during the 1980s, very few communities relied on sirens to warn the public. Other methods such as door-to-door warning, police/fire vehicle public-address systems, and tone alert radios were effectively used to warn the public. It is therefore recommended that you develop an adequate warning system based on your facility's risks. Do not assume that sirens are mandated or necessary simply because they are discussed in NRT guidance. Sirens are very effective notification and warning devices; however, there may be other methods that may be just as effective.

Another method mentioned in the NRT guidance is the Emergency Broadcast System (EBS). While this system can be effectively used for providing the public with emergency information, it should never be viewed as the sole means of notifying the public. Use the EBS as one part of a comprehensive warning system. The EBS can be effective as a warning mechanism only for those people watching television or listening to the radio. Obviously, additional methods are necessary. If your community is going to rely on the EBS as part of its overall warning system, carefully review all aspects of the state and local systems to determine their adequacy, and then test them frequently to ensure that they work as designed. It is generally acknowledged that the EBS does not receive the attention it deserves and therefore does not work as planned.

Section 15 recommends the development of a public information program:

15.0 PUBLIC INFORMATION AND COMMUNITY RELATIONS
The Plan should describe the methods used by local governments, prior to emergencies, for educating the public about possible emergencies and planned protective measures; describe the methods for keeping the public informed during an emergency situation, including a list of all radio, TV, and press contacts; and describe any related public information activities of covered facilities.

Pre-emergency public education/information programs have been proven effective in communicating risk to the public and for disseminating information during emergencies. For this reason, your facility should assist in developing these programs.

It is important to note that researchers have yet to determine what, if any, effect public information/education programs have on improving the effectiveness of public protective actions or public response during emergencies. Therefore it is recommended that you invest in risk communication programs in advance, and develop a strong capability to disseminate information during emergencies, but not to spend a lot of effort and money on public information brochures, and the like, that contain emergency preparation material, such as evacuation maps.

The NRT correctly recommends establishing an effective emergency resource management program:

16.0 RESOURCE MANAGEMENT
The Plan shall include a description of emergency equipment and facilities in the community at each facility in the community subject to the requirements of this subtitle and an identification of the persons responsible for such equipment and facilities. It should list personnel resources available for emergency response by major categories, including governmental, volunteer, and private sectors; describe the types, quantities, capabilities and locations of emergency response equipment available to the local emergency response units, including fire, police and emergency medical response units. [Categories of equipment should include that for transportation, communications, monitoring and detection, containment, decontamination, removal, and cleanup].

It should list the emergency response equipment available to each of the covered facilities and describe them in the same way as community equipment is described and describe the emergency operating centers or other facilities available to the local community and the facility emergency coordinators and other response coordinators, such as the incident commanders; describe the response facilities available to each covered facility and the conditions under which they are to be used in support of local responders; describe any significant resource shortfalls and any mutual support agreements with other jurisdictions whereby the district might increase its capabilities in an emergency; describe procedures for securing assistance from federal agencies and their emergency support contractors; describe emergency response capabilities and expertise in the private sector that might be available to assist both local responders and facilities and transportation companies in emergencies.

Developing an effective emergency resource management program requires a lot of time to research and identify all community resources and capabilities. It also requires an effective database management system. Any assistance you can provide in researching and identifying local resources would be of tremendous value to local planners. Developing a computerized database of this information would also be useful.

Facilities should assist local health care providers with assistance in preparing for emergencies that could occur at your site. The NRT recommends:

17.0 HEALTH AND MEDICAL

The Plan shall include methods and procedures to be followed by facility owners, operators and local emergency and medical personnel to respond to any release of such substances. The Plan should indicate the procedures for summoning emergency medical personnel and describe the procedures for emergency medical services, including first aid, triage, ambulance service, and emergency medical care, using both the resources available within the district and those that can be secured in neighboring districts; describe the procedures to be followed in decontamination of exposed persons; describe protective action measures designed to provide for sanitation, food, water supplies, and safe reentry of persons to the accident area.

It should describe procedures for conducting health assessments upon which to base protective action decisions and the provisions for emergency mental health care. The plan should describe the capacity of the emergency medical facilities, equipment and personnel available within the district; describe the level and types of capabilities in the district to deal with the medical problems created by exposure to extremely hazardous substances; and indicate mutual aid agreements with other communities to provide backup emergency medical personnel and equipment.

Providing an industrial hygienist or qualified medical specialist to assist in the development of this procedure is recommended.

The importance of protecting the health and safety of public response personnel is evident by the following NRT requirements:

18.0 RESPONSE PERSONNEL SAFETY

The Plan should describe procedures for assuring the safety of response personnel during an emergency response. [Just prior to publication of NRT-1, the Occupational Safety and Health Administration (OSHA) published proposed rules (29 CFR Part 1910.120) to provide more definitive requirements to plan for emergency response personnel safety. If the LEPC plans include a section on this function, the plan elements listed in the OSHA regulation should be used].

Your facility could help in this task by sponsoring joint training programs between facility personnel and government personnel. You could also provide funds for sending a few members of the local fire department to special hazmat or related schools.

The development of public protection plans is of utmost importance and is often a central point of LEPC discussions. The NRT mandates the following:

19.0 PERSONAL PROTECTION OF CITIZENS/INDOOR PROTECTION
The Plan should identify the decision-making process, including the decision-making authority for indoor protection; describe methods in place in the community and/or each of the covered facilities for determining the areas likely to be affected by a release, including methods to predict the speed, direction, and concentration of plumes resulting from airborne releases, and methods for modeling vapor-cloud dispersion as well as methods to monitor the release and concentrations in real time; describe the system for warning the public and advising on indoor protective measures; describe the roles and activities of covered facilities in recommending indoor protective measures; and describe the methods of indoor protection that would be recommended for citizens, including provisions for shutting off ventilation systems.

It should indicate the conditions under which such protection would be recommended, including the decision-making criteria; describe methods for determining the private and public property that may be in the affected areas and the nature of the impact of the release on this property; describe the capabilities to determine when indoor protection is no longer appropriate or required; describe the methods for educating the public on indoor protective measures.

To assist community planners in developing this section, you could provide information on how your facility determines areas likely to be affected by a release and how it models dispersion.

Section 20 of NRT-1 which concerns evacuation, is discussed in chapter 11.

Other areas of concern which should be addressed in local plans include: fire and rescue; law enforcement; incident assessment; human services, public works, spill containment; incident follow-up; any plan updates:

21.0 FIRE AND RESCUE
The Plan should list the major tasks to be performed by firefighters in coping with HAZMAT incidents; and list all major fire and rescue functions of covered facilities that complement or support local fire and rescue units. It should describe

the chain of command among firefighters, including the command structure when more than one jurisdiction is involved; list available support systems, e.g., protective equipment and emergency response guides, DOT Emergency Response Guidebook, mutual aid agreements, and good-Samaritan provisions; and list and describe any HAZMAT teams in the district.

You should help community planners develop procedures that indicate how fire units would interact with plant emergency brigades.

22.0 LAW ENFORCEMENT
The Plan should list the major law enforcement tasks related to HAZMAT incidents, including those related to security for the accident site and for evacuation activities and describe the chain of command for law enforcement officials; and list the locations of control points for the performance of tasks, with the appropriate maps.

You should help community planners develop procedures that indicate how police units would interact with plant emergency brigade and security personnel.

23.0 ONGOING INCIDENT ASSESSMENT
The Plan should describe methods and capabilities of both local response organizations and facilities for monitoring releases, including sampling around the site and the provisions for environmental assessments, biological monitoring the contamination surveys. It should state the responsibilities and methods of persons responsible for monitoring the size, concentration, and migration of leaks, spills, and releases.

You should provide community planners with information on the plant's accident assessment system.

24.0 HUMAN SERVICES
The Plan should list the agencies responsible for providing emergency human services, e.g., food, shelter, and clothing; and describe the major human services tasks of each agency.

25.0 PUBLIC WORKS
The Plan should list all major tasks to be performed by the public works department in a HAZMAT incident.

26.0 TECHNIQUES FOR SPILL CONTAINMENT AND CLEANUP
The Plan should describe the major containment and mitigation activities for all major types of HAZMAT incidents; and

describe methods to restore the surrounding environment, including natural resource areas to pre-emergency conditions. It should explain the allocation of responsibilities between local authorities and covered facilities and responsible parties for these activities and state the cleanup and disposal services to be provided by the responsible parties and/or the local community. It should also list cleanup material and equipment available within the district; and describe the capabilities of cleanup personnel;

The plan should outline provisions for long-term site control. It should list the locations of approved disposal sites; and describe applicable regulations governing disposal of hazardous materials in the district.

You should provide community planners with information on techniques for spill containment and cleanup for chemicals at your site. You could also provide funding for local firefighters to attend hazmat training seminars.

27.0 DOCUMENTATION AND INVESTIGATIVE FOLLOW-UP
The Plan should list all reports required in the district and all offices and agencies that are responsible for preparing them following a release and describe the methods of evaluating responses and identify persons responsible for evaluations; and provisions for cost recovery.

28.0 PROCEDURES FOR TESTING AND UPDATING THE PLAN
The Plan should include methods and schedules for exercising the emergency plan. It should describe the nature of exercises that the LEPC intends to conduct to test the adequacy of the plan; the frequency of such exercises, by type; contain an exercise schedule for the current year and for future years; describe the procedures by which it will evaluate performance in the exercise, make revisions to plans and correct deficiencies in response capabilities. It should also explain the role of covered facilities or transportation companies in these exercises.

Finally, section 29 of NRT-1 discusses the need for training:

29.0 TRAINING
The Plan shall describe training requirements for LEPC members and all emergency planners within the district; including training requirements for all the major categories of response personnel responsible for implementing the plan, including types of courses and the number of hours. It should list and

describe training programs to support these requirements, including all training to be provided by the community, state and federal agencies and the private sector; and contain a schedule of training activities for the current year and for the following three years.

Your company could provide opportunities for joint training and could provide funding for training local response personnel at specialized schools and seminars.

11

Community Evacuation Planning

Earlier chapters outlined the nature of various natural and technological hazards and their potential impact on humans. People must
take action to protect themselves from such hazards and the two
most basic methods of protection include sheltering and evacuation.
While the sheltering option is relatively simple to accomplish,
evacuations pose difficult problems and therefore must be addressed
in detail in an emergency preparedness program.

Population evacuation is not new. For centuries, cities and towns
have been evacuated to escape natural disasters and invading armies.
Most evacuations of the past have been spontaneous and disorga-
nized. But because today's population centers are larger and more
densely populated than in the past, we must have detailed plans for
evacuation so that resources can be effectively utilized and evacuations can be successfully executed. While public evacuation planning is clearly the responsibility of community planners, industrial
emergency planners should provide some assistance. This chapter
provides some basic information about evacuation planning.

EVACUATION FACTORS

Past research shows that, on the average, the United States has
enough vehicles and fuel to support massive evacuations of large risk
areas. In many locations, however, road capacity is limited, so that
careful planning and scheduling along with continuous monitoring
and control is needed to avoid severe traffic tie-ups. The key factors
in large-scale evacuations are:

1. Vehicles
2. Fuel
3. Road capacity

The availability of privately-owned vehicles is generally sufficient to support mass evacuations. Most families have access to a vehicle if sudden evacuations are mandated. Nationwide, 80% of all evacuees will evacuate an area in automobiles. Residents of large urban areas, however, do not always have access to private vehicles, and therefore, plans for the utilization of local bus fleets and school buses to evacuate autoless residents are necessary. In general, the availability of vehicles always exceeds demand; the primary problem in a large evacuation is not a lack of vehicles but rather managing and organizing the evacuation.

Research has shown that most cars have enough fuel to allow for evacuation. The demand for fuel may rise in host areas after the evacuation is completed, as a greater number of individuals will want to purchase fuel than is normally the case. This should pose a significant problem only if a large number of people are evacuated to an area with limited resources.

The road capacity of any area could be taxed by an evacuation. The extent, however, will be determined by the number of people being evacuated over a certain period of time. Careful planning, combined with continuous monitoring and broadcasting of traffic conditions will alleviate excessive bottlenecks and other local road network problems.

The NRT-1 section concerning evacuation is as follows:

20.0 Personal Protective Measures/Evacuation Procedures
The Plan shall include evacuation plans, including provision for a precautionary evacuation and alternative traffic routes. It should: describe methods to be utilized in evacuation, including methods for assisting the movement of mobility-impaired persons and in the evacuation of schools, hospitals, prisons and other facilities describe evacuation routes, including primary and alternative routes; [These may be either established routes for the community or special routes appropriate to the location of facilities]; describe potential conditions requiring evacuation, i.e., the types of accidental releases and spills that may require evacuation; describe evacuation zones and distances and the basis for their determination; [These should be related to the location of facilities and transportation routes and the potential pathways of exposure]; describe procedures for ac-

complishing a precautionary evacuation for special populations; and should describe the role of any covered facilities in evacuation decision making;

It should list facilities for the provision of mass care to a relocated population, including food, shelter and medical care; describe evacuation responsibilities of various governmental agencies and supporting private organizations, such as the Red Cross, and the chain of command among them; describe procedures for providing security for the evacuation, for evacuees and of the evacuated areas; describe methods for managing the flow of traffic along evacuation routes and for keeping people from entering and reentering threatened areas, including maps with traffic and other control points; describe the provisions for managing an orderly return of people to the evacuated area; and discuss the authority for ordering or recommending evacuation, including the person(s) authorized to order an evacuation.

PLANNING APPROACH

When planning for evacuations, especially complex evacuations, you can consult the Federal Emergency Management Agency (FEMA) and the Nuclear Regulatory Commission (NRC). Both of these government agencies have published literature about large-scale evacuation planning. For a basic overview of planning for evacuations from hazards associated with fixed-site facilities, consider the following steps:

1. Determine the worst case scenario that would cause an evacuation of neighboring areas. These areas can be referred to as risk areas.

2. Obtain the population data (including day, night, and seasonal) for all risk areas. Also determine the availability of vehicles for the potential evacuees and which roads are available for evacuation.

3. Identify potential host areas. For short-term evacuations, such areas could simply be nearby schools, churches, auditoriums, and other facilities with enough space, bathrooms, and kitchen facilities to support the evacuees. If large numbers of evacuees are expected to require shelter for a longer duration, facilities such as hotels or motels might be necessary. (**Note:** In most communities, the local Red Cross chapter and civil-defense authorities are responsible for caring for evacuees. Therefore, it

is recommended that evacuation plans be developed in conjunction with these local officials.)

4. Identify all possible evacuation routes between point of origin and the host area. By talking to local highway officials, determine the peak capacity of these roads and identify possible bottleneck areas and major intersections. Select the best possible routes to handle the expected number of evacuees.

5. Develop warning plans, traffic control plans, monitoring plans, and perimeter control plans. Traffic flow and perimeter control is usually conducted under the direction of law-enforcement officials.

6. Develop plans for evacuating those residents who may not have access to a vehicle at all times of the day and for handicapped individuals. A survey can be used to determine who is autoless or handicapped; however, this method is often inaccurate, and the results are quickly outdated. A better idea might be to have a certain number of buses and bus drivers ready to move through the risk areas to pick up persons who need transportation.

7. Produce a written plan (or annex to the local emergency operations plan) that incorporates all of the preceding steps, and exercise the plan to determine its feasibility.

It is important to reiterate that planning for public evacuations is the responsibility of local officials, and therefore, private industry officials should only assist in this effort, not dominate it.

12

Emergency Training Programs

OVERVIEW

A good emergency plan specifies what is to be done during an emergency, who is to do it, and how it is to be done; and the only way that personnel can understand their emergency responsibilities is with good training. All emergency training should be based on the specific hazards addressed by the contingency plan and on the response duties of each individual. Before developing a training program, an assessment should be undertaken to pinpoint specific needs and areas requiring attention. Three things must be determined in this assessment: (1) the responsibilities of each individual under the contingency plan, (2) the specific tasks necessary to perform those responsibilities, and (3) the skills and knowledge needed by the person to perform those tasks. This is accomplished through a detailed review of the emergency plan and related procedures, interviews with the individuals about their assigned emergency responsibilities, and an analysis of other types of training provided for the employees.

For example, a member of an organization's emergency brigade might be required to respond to a hazardous materials spill. Some of the skills and knowledge the person would require in order to effectively respond would include a knowledge of the hazard, first-aid measures, protective clothing and equipment, and spill-control techniques, among others. The portion of the training program dealing with protective clothing could include a combination of classroom instruction and field demonstration of the types of available clothing and equipment, levels of protection, criteria for selection, maintenance, and finally, a practice drill utilizing the clothing and equipment.

A sound training program can then be developed based on specific learning objectives identified in this analysis. Various methods of training can be employed to accomplish these objectives and may include classroom instruction, periodic staff meetings, or field demonstrations and drills. The training program should have very specific guidelines for the type and frequency of training for each individual with emergency response functions. Students should be tested following each training session, because it is only through the use of testing devices that you can be assured of the employee's ability to understand and perform required tasks. The adequacy of training should be reviewed periodically following drills, exercises, or actual emergencies to ensure that employees learn from past experiences. Also, training programs should be modified as necessary to reflect changes in the plan or new hazards at the facility.

Finally, whereas it is quite probable that the facility's staff will at one time or another interface with off-site authorities, it is important and wise to coordinate certain training with government response agencies. For example, joint training should be conducted between the firm's fire brigade and the community's fire department; security staff should train with the police department; and staff industrial hygienists should train with the community's health or environmental protection agency. Figure 12–1 illustrates the development process for the emergency training program.

Based on the fact that there are many regulations requiring employee emergency training and that it is important to develop an effective emergency response program, the facility's human resources department or emergency management coordinator should take the lead in developing the emergency training program. The following provides a systematic approach to developing an emergency training program. This approach, in its generic form, is used by thousands of professional trainers in developing a number of corporate training programs.

SYSTEMATIC APPROACH TO TRAINING

Job/Task Analysis

The first step in performing a systematic analysis is to determine the emergency job performance requirements, the training needs of job incumbents, and prerequisite qualifications of the specific emergency jobs for which training will be given. The trainer should systematically identify and analyze all job functions that are impor-

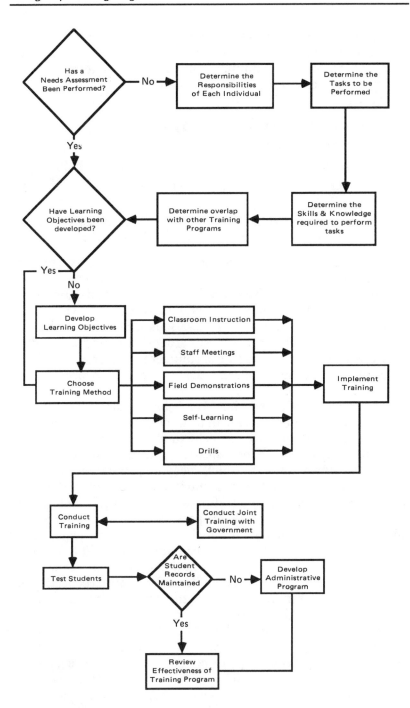

Figure 12–1: Training program development

tant to effective emergency response performance. Following this analysis, the trainer should develop a "job/task profile" for each emergency position's performance requirements in terms of duties and tasks. The basic format of the job/task profile should include the following:

1. **Mission.** The overall goal of the position.
2. **Major duties.** Fully describe the job in terms of duties.
3. **Tasks.** State the tasks to be performed under each duty.
4. **Task description.** State exactly what is done by the job incumbent.
5. **Team versus individual.** Distinguish between tasks performed by an individual and tasks performed by a team.

Figure 12–2 is an example of such a job/task profile for a member of an emergency response team.

Once this profile has been developed, all identified duties, tasks, and associated task information should be verified. This can be accomplished in accordance with various established techniques (e.g., surveys, interviews, consensus groups) or more commonly by collecting verification data from a sufficient number of knowledgeable job incumbents to ensure that the profile is accurate and comprehensive.

At this point, the trainer should probably be able to determine the qualification requirements for all tasks determined to be important to emergency task performance. The trainer should then determine the training needs of job incumbents and initial trainees. The first step is to identify tasks for which adequate performance cannot be ensured by methods other than training (e.g., by job performance aids). Next, the trainer should estimate the job skills, knowledge, and previous training and experience possessed by initial trainees and identify associated tasks for which training is not required. For example, an emergency response team member might have to perform the task or task element of turning off lights. It can be safely assumed that all employees have at least the minimal education and experience necessary to accomplish this task and it therefore need not be included in the training program.

The trainer should then select tasks for training and/or periodic retraining, basing the principal selection criteria on the importance of the task to emergency response performance (e.g., consequences of inadequate performance) and basing secondary selection criteria on factors indirectly affecting emergency response performance (e.g., probability of error, existing performance deficiency, frequency of performance, team coordination).

```
┌─────────────────────────────────────────────────────────┐
│                 Emergency Job/Task Profile                │
├─────────────────────────────────────────────────────────┤
│ Position: Emergency Response Team Member                  │
├─────────────────────────────────────────────────────────┤
│ Mission: Team members are responsible for responding to all│
│          emergencies arising at the plant - - including fire &│
│          chemical spills                                  │
└─────────────────────────────────────────────────────────┘
```

Major Duties: ─────────────────────────────────────
   1. Respond to plant fires
   2. Respond to chemical spills
   (etc.)

Tasks: ──
   1. Respond to plant fires
       • use fire extinguishers
       • handle hoses
       • identify fire cause
       • *(etc.)*
   2. Respond to chemical spills
         (continued on reverse)

Task Description:	**Type**
1. Use fire extinguishers	
• check for level	Individual
• check extinguisher type	Individual
2. Handle hoses	
(etc.)	

Figure 12-2: Sample job/task profile

It is important to consider the specific skills and conditions that relate to each assigned task, as can be accomplished by the following:

1. Determine the necessary steps (task elements), and their sequence, required for the successful performance of each task.

2. Identify the conditions under which the task (or task element) is performed.

3. Identify the initiating and terminating cues for task (or task element) performance.

4. Specify the standards to which the task (or task element) must be performed.
5. Identify the skill(s) and type of skill(s) (e.g., physical or mental) required in the performance of each task and task element.
6. Identify the knowledge necessary to support each identified skill.
7. Continue the task analysis to the level of detail at which the identified skills and knowledge match the skills and knowledge assumed for trainees meeting the facility's minimum selection criteria (e.g., high school graduate).

As with the job/task profile, all identified skills and knowledge for each task should be verified. Again, this can be done in accordance with established techniques (e.g., surveys, interviews, consensus groups) or by collecting verification data from a sufficient number of knowledgeable job incumbents.

Developing Learning Objectives

Based on the job/task analysis, learning objectives can be identified that describe desired trainee performance after training. Two types of learning objectives must be identified: (1) terminal learning objectives and (2) enabling learning objectives.

Terminal learning objectives describe exactly what the trainee will be able to do at the conclusion of training in terms of measurable performance. For example:

1. Specifically stating the observable action or behavior the trainee(s) is (are) to exhibit at the conclusion of training.
2. Specifically stating the conditions under which the action should occur.
3. Specifically stating the standard and criterion that must be met for adequate performance.

Then translate task elements, skills, knowledge, and related information associated with each terminal learning objective into **enabling learning objectives** by:

1. Specifically stating behavior, conditions, and standards.
2. Directly relating each enabling learning objective to a specific terminal learning objective(s).
3. Distinguishing between knowledge and different types of skills (e.g., physical and mental).

Curriculum Design

The design of the emergency training program curriculum should be based on the specified learning objectives selected for training. All aspects of instruction (e.g., instructional methods, tests, media, and training program management) should be determined systematically using the learning objectives as the primary basis for all decisions.

Sequencing the learning objectives. It is important to identify those learning objectives that must be mastered before certain other learning objectives can be reasonably achieved. Then identify those learning objectives for which, because of their similarity, the mastery of one makes mastery of the others easier. Then identify those learning objectives that can be mastered independently of one another. Use these relationships to organize all the learning objectives into the most logical sequence in which they should be mastered.

Systematically identifying training methods. The trainer should identify the instructional method(s) (e.g., lecture, on-the-job training, simulation, self-study, team training) and testing method(s) (e.g., written, oral, performance) based on: (1) learning-objective performance requirements, (2) instructional requirements (e.g., initial training, retraining), and (3) requirements for trainee/instructor interaction.

The trainer should also identify instructional media based on: (1) the type of performance specified by the learning objectives (e.g., use of rule, verbal communication), (2) the type of learning required (e.g., mental skills, physical skill), and (3) instructional media characteristics required to support the type of performance and the type of learning (e.g., initiating cues associated with a specific learning objective may require dynamic visual presentation for effective mastery learning; a medium such as video tape may be selected to satisfy this requirement).

Curriculum Development

This step involves the development of the training program material required to support the learning objectives. This includes preparation of standardized instructor lesson plans, classroom visual aids, student text material, and so on.

Organizing the instructional content involves the following:

1. Organizing the instructional content for initial training, based on the learning objectives sequence.

2. Providing for sufficient rehearsal (practice) and reinforcement (performance feedback) of previously mastered learning objectives as new instruction content is introduced, to ensure retention and integration of the associated skills.

3. Organizing instructional content into parts (e.g., modules, segments, phases, lessons) in accordance with the selected instructional methods, media, and scheduling considerations.

4. Selecting instructional content for periodic retraining, based on the tasks identified for training.

5. Organizing the instructional content for retraining, in accordance with the same considerations as for initial training.

Preparing instructional media involves the following:

1. Preparing lesson plans and supporting instructional information to include:

   • a summary overview of each subdivision of instruction

   • the instructor's role and actions

   • instructor/trainee references

   • instructional methods and media to be used

   • the sequence of events during training

   • a description of the interrelationship of instructional content, job requirements, and the training program

   • trainee learning objectives

   • criteria for successful trainee performance

2. Evaluating the capacity of existing instructional media (e.g., lesson plans, student text material, visuals) to support the achievement of identified learning objectives; modifying and incorporating such media as required by the training program design.

3. Developing new instructional media as required to support the training program design, using established methods.

Trainee Evaluation

The trainer should develop appropriate tests from learning objectives according to the performance standards and evaluation criteria delineated and should specify necessary procedures and guidelines for administering tests to maximize consistency and job relevance. Tests should be developed for both terminal and enabling learning objectives. Each test should require the demonstration of knowledge and/or skills directly related to the associated learning objectives and underlying tasks. Define the performance standard and criterion for each test in order to clearly distinguish between adequate and inadequate learning objective mastery. Provide a scoring key and directions specifying the necessary procedure(s) and resources for giving each test. The trainer should systematically analyze test results and provide effective performance feedback to the trainee. Not only will this analysis help to improve the trainees' deficiencies, it will help the trainer identify training program deficiencies for subsequent corrective program modifications.

Program Implementation

The trainer should prepare and implement a training program management plan that provides for organizing, controlling, and evaluating the delivery of emergency response training to the trainees. The plan should clearly delineate the assignment and definition of the training program's administration, instructorship, and support responsibilities. It should delineate the trainee management strategy, including program entry and completion criteria, and procedures for identifying/controlling marginal trainees. This is especially important, since during an emergency a marginal trainee could get seriously hurt or hurt someone else.

The plan should specify instructional facilities (e.g., building, classroom laboratory, equipment) and instructional media requirements. Some emergency training may have to take place at a special facility, such as a county's fire-and-hazardous-materials training academy. It should specify the training program evaluation plan and provide for the delivery of systematic performance feedback to the trainee throughout the training program.

It is important to note that training should be conducted in accordance with the training program management plan. It would be a waste of vast resources to develop a strong emergency training program and then not be able to follow through with the plan. Also, only qualified personnel should conduct the emergency training. Instructor performance requirements and technical qualifications should be established to ensure effective training.

Program Revision

A plan for the systematic and regularly scheduled evaluation of the effectiveness of the training program should be developed and implemented as indicated by performance of trainees in the actual emergencies or exercises setting. Revision and upgrading of the training program should be based on such evaluations. The need for changes might be the result of poor design or development, changes in procedure, or other external factors.

EXISTING PROGRAMS

In addition to developing a custom emergency training program, companies can take advantage of existing courses and higher-education programs to fulfill their needs. Hazardous materials response courses and emergency management courses are offered by a number of professional trade associations, private for-profit firms, private not-for-profit firms, and colleges and universities.

Colleges and universities are only now starting to develop programs to meet the needs of industry in the hazardous materials emergency response arena. There are over a half dozen universities in the United States that offer such programs now that did not just five years ago. Universities that offer hazardous materials programs include the University of Virginia, the University of North Carolina, State University of New York, Johns Hopkins University, and Tufts University. These universities offer one or more of the following subjects, all of which have been identified as lacking in most engineers who have completed traditional environmental engineering programs (or related fields): environmental law, hydrology, hazardous materials transportation, treatment technology, laboratory testing and analysis, and waste minimization. North Texas State University was the first to offer degree programs in emergency planning and response, and the University of Wisconsin has a certification program in emergency management, although it is generally geared towards international affairs. These programs generally fill industry's long-term needs by training young people for the future, although some colleges offer seminars and continuing education for professionals already in industry.

For many companies, emergency training needs may be better resolved in the short term by taking advantage of specialized training courses. Several professional/trade associations offer short courses or manuals related to emergency planning, emergency response, hazardous materials and the like. Among them are the following:

- The American Society of Safety Engineers
- The Chlorine Institute
- Chemical Manufacturers Association

Several other associations offer programs as well. If you are a member of a professional society, check with its training group to determine what might be available. Often, these courses are geared to your specific profession. It is likely that certain industry groups have already developed or modified programs to meet the specific emergency planning and response requirements of your particular industry. There are a few programs for hazardous materials responders that are geared towards the people already working at the trade. Two of these are the following:

- The Association of American Railroads—Hazardous Materials Training Transportation
- The Environmental Protection Administration's Emergency Response Team Training Program.

Other, private programs are also available.

13

Designing Emergency Drills and Exercises

PURPOSE AND OBJECTIVES

Drills and exercises have two basic functions, namely training and testing. While exercises do provide an effective means of training in response procedures and in sharpening individual technical skills, their primary purpose is to test the adequacy of the emergency management system and to ensure that all response elements are fully capable of managing any emergency situation.

Because drills and exercises simulate actual emergency situations, they are the best means of accomplishing the following goals and objectives:

1. To reveal weaknesses in the plans and procedures before emergencies occur.
2. To identify deficiencies in resources (both in manpower and equipment).
3. To improve the level of coordination among various response personnel, departments, and agencies.
4. To gain public recognition and confidence in the plant's ability to manage emergency situations.
5. To improve the proficiency and confidences of emergency response personnel.
6. To clarify each individual's role and areas of responsibility.
7. To increase the government's and community's cooperation with the facility's emergency planning effort.
8. To enhance overall emergency response capabilities.

TYPES OF DRILLS AND EXERCISES

The four types of drills and exercises used by a plant to test its ability to deal with emergencies are: (1) orientation exercises, (2) tabletop exercises, (3) functional drills, and (4) full-scale exercises. Each of these are designed to evaluate individuals' responses to various degrees of simulated emergency conditions in order to test the adequacy of training equipment and procedures.

Orientation exercises, or seminars, are used to familiarize people with new plans and procedures and to provide for an initial review of the plan by key personnel. Orientation exercises are used during the initial implementation of a new plan, when major plan revisions are made, or when there are changes in key personnel. The orientation exercises employ only a low degree of realism. They are presented initially to all response personnel and should include government and community organizations as well. Thereafter, only those personnel affected by changes or revisions would be involved. Orientation exercises require approximately two weeks of preparation time. They require the use of a conference or training room large enough to accommodate the number of individuals involved, and they may require the preparation of special displays and materials.

Tabletop exercises provide a walk-through of a simulated emergency situation without the pressures and time restraints that would be present during an actual emergency. Tabletop exercises are similar to orientation exercises except that their purpose is to evaluate plans and procedures, develop problem solving and interorganizational coordination skills, evaluate the effects of new hazards and operations on the plan, and aid in training. A tabletop exercise involves the use of simulated events in the narrative form. Participants discuss possible solutions and responses to these events, utilizing the procedures outlined in the emergency plan. A degree of realism can be created by introducing simulated emergency events to participants through a series of messages or problems. These messages or problems would be distributed at intervals to simulate real messages or events that would be occurring in an actual emergency situation.

Approximately one month lead time is required for the development of a tabletop exercise. A general-purpose room large enough for all of the participants and for simulation materials (maps, charts, etc.) is required. Special displays and materials may be necessary as well as the use of some emergency response equipment (e.g., computers). Prior to the tabletop exercise, all participants should have undergone the initial training required to perform their assigned tasks, and an orientation exercise. Government, community, and mutual aid organizations should be encouraged to participate in tabletop exercises, whenever appropriate.

Functional drills are designed to test or evaluate the capability of individual components of the response organization or portions of the plan (e.g., an exercise limited to a test of the fire response team). Functional drills are conducted to reinforce established procedures, test personnel and equipment readiness, and evaluate the effectiveness of training. Functional drills require participants to respond to realistically simulated emergency conditions. Messages used to control the exercise should be delivered using the same methods (radio, telephone, etc.) and form as would be expected during a real emergency. Mock damage, simulated injured employees, simulated spills, and the use of smoke generators or other similar devices should be used whenever possible, to create the highest level of realism that can be safely achieved.

Because of the high degree of realism, extensive time is required for the preparation of functional drills. Up to three months lead time should be allowed to develop the necessary simulation materials and plans. All participants should be fully trained in and participate in a tabletop exercise prior to participating in a functional drill.

Full-scale exercises are intended to provide a test of the entire emergency response system. While functional drills test the capabilities of individual plan elements of a response function, a full-scale exercise is intended to test the entire plan and response organization. A full-scale exercise should involve many elements of the emergency plan, especially those areas involving coordination of plant activities with government, community, and mutual aid organizations. In addition to involving a wider segment of the emergency plan, full-scale exercises involve placing considerably more stress on exercise participants and are longer in duration than other forms of exercises.

Full-scale exercises require an experienced planning staff with representation from all involved organizations. Six months lead time for planning and development is required. Since a full-scale exercise often involves other organizations, during the development of the exercise's scenario and schedule, planning must consider the needs of those organizations.

CYCLICAL NATURE OF EXERCISES

The four types of exercises are intended to be used in an ongoing cycle of progressively complex exercises. It is based on the concept that to walk, one must first crawl. Each step in this cycle is more complex than the previous one and requires more preparation time, more personnel, more planning, and a higher degree of realism. The cycle begins with the training of personnel, then to orientation exercises, through tabletop exercises, to functional drills, and finally to full-scale exercises. In other words, an orientation exercise, a tabletop exercise, and functional drills should be conducted for each major plan component before the full-scale exercise is conducted.

In subsequent years, a schedule of drills and exercises should be established consisting of an annual full-scale exercise preceded by at least one tabletop exercise and several functional drills. The actual number of orientation, tabletop, or functional drills that are necessary each year will be based on the facility's analysis of its needs. However, the following critical response functions and corresponding elements of the plan should be tested using one form of exercise or another on a yearly or more frequent basis:

- full-facility evacuation, including a test of the alarm/notification procedures and employee accountability
- interior structural fire suppression
- emergency first aid
- search and rescue techniques

The emergency coordinator should establish a tentative schedule for an exercise program each year. While this schedule is likely to be modified due to business concerns, it will set a goal that will help ensure that an adequate exercise program is implemented.

PREPARATION AND PLANNING

Preparation is the key to success! As previously stated, the degree of planning and time required for preparation of the various types of exercises varies considerably. However, some basic steps common to exercise preparation include the following:

- determining the purpose (i.e., needs analysis)
- identifying available resources and the capability to conduct the exercise
- planning the exercise (including staff assignment, refining objectives, determining scope, and identifying necessary logistical needs)
- developing the exercise scenario and simulation materials
- assigning exercise staff functions (controllers, simulators, evaluators)

Although the amount of preparation for each exercise will vary greatly, the basic concepts just discussed need to be considered if the exercise is to be successful. As an aid to ensuring the adequacy of planning, refer to the "Exercise Planning and Scheduling Checklist" in figure 13-1. The emergency coordinator should begin the planning process by assigning target dates for the completion of each item. Each item should be checked off and the actual completion date filled in as each item is finished.

Exercise Planning and Scheduling Checklist		
Item	Scheduled	Completed
_____ Conduct needs analysis		
_____ Conduct resource analysis		
_____ Determine planning-staff needs/appoiint staff		
_____ Determine plan element(s) to be exercises		
_____ Define exercise scope		
_____ Select exercise type		
_____ Obtain participants' committments		
_____ Determine costs/liabilities		
_____ Develop statement of purpose		
_____ Refine objectives/expected actions		
_____ Develop scenario narrative		
_____ Develop MSEL (master sequence of events list)		
_____ Develop messages/problems		
_____ Determine facility/equipment requirements		
_____ Determine communications requirements		
_____ Develop exercise simulation materiials		
_____ Determine exercise-staffing needs		
_____ Conduct staff training		
_____ Develop evaluation material		
_____ Conduct pre-exercise briefing		
_____ Begin exercise		

Figure 13–1: Sample exercise planning checklist

Needs Analysis

The starting point in the planning process is to identify the reasons for conducting the exercise. This is accomplished through **needs analysis**. The requirement for conducting an exercise is identified through an analysis of the emergency plan to identify those critical elements where there is a need to confirm the element's credibility. This need can exist because the plan element has been recently implemented or revised, because there have been changes in the facility or personnel, or because it has not been recently exercised or used.

To facilitate the needs analysis, refer to figure 13–2. The left-hand column lists all major emergency plan elements or response functions. The primary factors in determining whether there is the need to conduct an exercise for those elements are shown in the other columns. These factors are:

1. **Newness.** Has the plan element been recently implemented?
2. **Revisions.** Have there been recent changes in the element's provisions?
3. **Change in personnel.** Have there been changes in key personnel responsible for implementing the element's provisions?
4. **Changes in facility or hazard.** Have there been any changes to the facility, manufacturing operations, or hazards that may have an effect on response capabilities?
5. **Date last used or exercised.** What is the date of the most recent use of the element in an exercise or actual emergency?

Any time you can answer yes to one of the first four factors, place a check mark in the appropriate space in its column. In the final column, record the appropriate date for each plan element.

Determining the order of priority for selecting the plans elements to be tested is subjective but emphasis should be based on the following:

1. Newness: Any new element should be tested.
2. Date last used: Plan elements that have not been tested in a long time should receive the highest priority.
3. Plan elements with multiple check marks should receive priority.
4. For elements for which only single check marks (other than "newness") apply, the order of priority is:

NEEDS ANALYSIS FORM					
Plan Element	Newness	Revision	Change in Personnel	Change in Facility or Hazard	Date Last Used or Exercised
Emergency Response Center					
Fire Response Team					
Spill Control Team					
Emergency Medical Team					
Environmental Monitoring Team					
Security Team					
Detection and Alerting					
Accident Assessment					
Emergency Communications					
Evacuation					
Off-site Coordination					
Emergency Public Information					
Transportation					
Resource Management					
Damage Assessment					
Cleanup and Salvage Operations					
Emergency Shutdown					
Hurricane Plan					
Bomb Threat					
Other: _____					

Figure 13–2: Sample needs analysis form

    a. revision

    b. change in personnel

    c. change in facility or hazard

It should not be interpreted that it is not important to test those areas not represented by the preceding guidelines.

Resource Analysis

The next step in the exercise planning process is to identify the resources available for performing the exercise. The questions outlined on the resource analysis form in figure 13–3 can help accomplish this. This analysis is not concerned with identifying the resources necessary to perform the actual response tasks, but rather with identifying those resources necessary to plan, prepare, and conduct the exercise. Defining these capabilities is a necessary first

---

Resource Analysis

Exercise Experience
Date of last exercise: _____ Type: _____
Major deficiencies identified: _____
_____
_____

Facilities
Are exercising-staging areas available within plant boundaries? _____
If so, what area? _____
_____

Are there any limitations to using the EOC during an exercise? _____
If so, what are they? _____
_____

Are meeting rooms available for use during an exercise? _____
Are off-site areas available? _____
If so, what areas? _____

Personnel
What personnel are available to assist in exercise planning? _____
_____
_____

Who is available to serve as:
controllers? _____
evaluators? _____
simulators? _____

Equipment
What equpment is available for use in exercises?
two-way radios: _____
telephones: _____
audio-visual equipment: _____
simulation equipment: _____
other: _____

Figure 13–3: Sample resource analysis form

step in selecting the type(s) of exercise, since the facility may not be able to conduct all exercise types with its present resources.

A primary objective of the resource analysis is to determine the availability of experienced personnel to help plan the exercise. For exercises involving on-site response by facility personnel, the plant's emergency planning coordinator should act as the exercise director. In a full-scale exercise, a co-exercise director may be appointed from an off-site organization, giving the emergency planning coordinator responsibility for facility planning and giving the co-exercise director responsibility for off-site planning. The exercise director(s) may be supported by a planning staff of individuals representing the various organizations and groups to be included in the exercise. Ensuring that each group or organization is represented on the team will help to create a coordinated effort. The function of this group is to: (1) determine exercise objectives, (2) tailor the exercise scenario, (3) develop the master sequence of events list (MSEL) and the exercise messages, and (4) assist in development of other exercise materials. For exercises involving only one department, a planning group may be unnecessary. In such a case, the emergency coordinator would assume all planning responsibilities and would request any assistance as necessary.

PLANNING THE EXERCISE

Once the needs analysis and resource analysis have been completed and a planning staff assigned, the actual planning process can begin. This involves the following steps:

- defining the scope
- selecting the exercise type
- determining costs and liabilities
- developing a statement of purpose
- refining objectives

Defining the Scope

By defining a scope, the basis for the exercise is determined. Defining the scope of the exercise involves analyzing six conditions relating to the facility and the emergency plan. The conditions are: (1) operations, (2) involved organizations, (3) personnel, (4) hazards, (5) geographical area, and (6) degree of realism.

Operations. Defining the scope of operations requires the identification of the specific response tasks that participants will be required to perform. While the needs analysis process has identified the overall activity to be exercised, it is necessary to define the specific tasks or operations that will be involved. For example, to determine the scope of an exercise to test fire suppression procedures, it is necessary to first determine what tasks may be involved, such as the use of fire extinguishers, communications with the EOC, and coordination with the local fire department.

Involved organizations. To further define scope, it is also necessary to identify the various organizations that will be expected to participate in the exercise. Once the operations have been determined, it is possible to identify all of the organizations that need to be involved. Using the previous example, the operational tasks to be used in the exercise require the involvement of three organizations: the fire response team, the EOC staff, and the local fire department. In planning some exercises, the organizations to be involved may not be as apparent as in this example. Therefore, the emergency plan should be reviewed to determine which groups have responsibility for certain tasks.

Personnel. By identifying the organizations that will participate in the exercise, specific personnel can also be identified. Not all personnel for each selected organization need be involved in each exercise.

Hazards. With regard to determining the type of hazard(s) involved, two factors should be considered:

1. The hazard(s) must be specific. For example, if a fire hazard is to be used in the exercise, it should not be stated that "a fire occurs" but rather "a solvent fire occurs in the warehouse area".

2. Determine the level of risk involved. This means defining both the probability of occurrence and the potential severity. The use of unrealistic hazards may diminish the effectiveness of the exercise.

Geographical area. The geographical area to be involved in the exercise should be one that is a logical place for the hazard to occur or for actual response actions to take place.

Degree of realism. The degree of realism refers to the amount of stress, complexity, and time pressure that the planning group desires to use. The decision as to the degree of realism must be made early in the planning process. The level of realism is often limited by physical restraints (such as the nonavailability of an area where a simulated fire can be set) and by budgetary restraints; however, the maximum degree of realism possible to meet the objectives of the exercise should be used.

Selecting the Exercise Type

With the completion of the preceding steps, it is possible to select the type of exercise to be used. When deciding on a type of exercise, the cycle of drills and exercises previously outlined should be taken into consideration. More complex exercises should always be preceded by the less complex. For example, before a functional drill is conducted, one or more tabletop exercises should be conducted; before a full-scale exercise is conducted, one or more functional drills should be completed. This progressive approach helps ensure that the complexity of an exercise does not exceed the participants' capability to perform their duties.

Each of the four types of drills and exercises have differences in purpose, degree of realism, scope, and resource needs. By comparing the requirements of the different exercise types to the purpose and scope of the exercise being planned, the selection of the best type of exercise is possible.

Orientation exercises are intended to be limited to the review of the plan following initial implementation, major revisions, or changes in key personnel.

Tabletop exercises are best used when the exercise's objective is limited to the development of problem solving, interorganizational coordination, and leadership skills, and when the number of personnel available to participate is limited, the degree of realism is limited, and the resources necessary to conduct it are minimal.

Functional drills should be used when technical skills, training, adequacy of equipment, or procedures are to be tested. They have the advantage of allowing a high degree of realism. Because functional drills limit the number of organizations, personnel, and operations involved by testing only one element of the emergency plan, the amount of resources necessary to conduct one is far less than for a full-scale exercise.

Full-scale exercises, because of their costs and lengthy preparation time, are limited to once per year. These exercises should be used to test situations requiring coordination of multiple groups and organizations.

Determining Costs and Liabilities

The period of early planning is also the time to address the issue of estimated costs and liabilities of the intended exercise. Costs will be incurred at every step in planning and conducting an exercise, and estimates should include the cost of personnel, equipment, supplies, and miscellaneous expenses (such as refreshments during exercises).

The question of who is to be responsible for various expenditures in exercises where off-site organizations are to be involved needs to be clearly resolved. Adequate funds for conducting exercises needs to be considered in the plant's annual budget to ensure that the exercise program will be carried out.

Liabilities need to be recognized and most often exist in functional drills and full-scale exercises, where the possibilities of personal injury or property damage is greatest. This is especially true when government and private groups other than your facility's personnel are involved. Insurance coverage should be checked before an exercise takes place to ensure that there is adequate coverage in case of an accident involving personnel or property during the exercise.

Developing a Statement of Purpose

An important step in the planning process is the development of a statement of purpose for the exercise. This statement must clearly and briefly state what is to be accomplished by the exercise. A purpose statement is written largely from the information obtained during the needs analysis and incorporates the information identified when the scope of operations and groups was defined. It should identify the following:

- element of the plan being tested
- operations involved
- organization(s) involved

A typical statement of purpose is as follows:

STATEMENT OF PURPOSE
The purpose of this exercise is to test the adequacy of response procedures to an interior structure fire by: (1) enacting the notification of the fire department, fire response team, and ECC [emergency control center] personnel, (2) recording the response times of these groups, (3) observing the communications procedures among these groups, and (4) assessing the coordination of efforts among the various groups.

The statement given identifies the **element of the plan being tested**: an interior structure fire response procedure. The **operations involved** are identified: notification of personnel, response by emergency personnel, communications, and coordination of the various groups. The **organizations involved** are identified: the fire department, the fire response team and the ECC staff.

Refining Objectives

The purpose statement just discussed is used as the basis for defining the specific objectives to be met by the exercise. These objectives are then used for defining the actions that are expected of the exercise participants. While the statement of purpose has defined what element of the plan is to be exercised, refining objectives leads to clear goals to be accomplished. Using the preceding example, some objectives for the portion of the exercise concerning notification procedures might be:

1. To assess the fire-reporting procedures from a line employee at the scene.

2. To assess the use of the plant alarm-system to warn of an emergency.

3. To assess the reliability of procedures used to contact the fire department.

Once objectives have been defined for all operations of the intended exercise, the next step is to define specific actions that would be expected of exercise participants to accomplish these goals. These are generated from the procedures contained in the emergency plan. These expected actions are important because they define what the exercise scenario must accomplish (what is necessary to trigger the action) and are the standards against which the participant's actual response will be measured. Three examples (using the objective 1, above) of several actions to be performed by a line employee might read:

1. Employee discovers fire and triggers alarm at nearest pullbox.

2. Employee warns employees in immediate area.

3. Supervisor is informed of nature, location, and extent of fire.

The actual process of refining objectives and defining the expected actions is relatively simple. By reviewing the appropriate section(s) of the emergency plan (identified by the statement of purpose) against the information developed by needs analysis, and with due consideration to scope and type of exercise, it it possible to identify exactly what must be tested and evaluated by the exercise. Specific procedures in the plan will identify those actions that are necessary for meeting exercise objectives.

14

Developing Exercise Materials

DEVELOPING SIMULATION MATERIALS

Once the exercise planning process is completed, it is necessary to develop the simulation materials necessary to achieve the desired results. A disaster scenario must be devised that simulates a real situation requiring the plan elements to be tested. From that scenario a series of realistic events and the means of introducing those events to exercise participants need to be created to trigger participant actions. This is accomplished through the creation of three basic documents:

- scenario narrative
- master sequence of events list (MSEL)
- messages and problems

Scenario Narrative

The scenario narrative is intended to set the stage for the simulated disaster. It should define the starting point of the exercise and briefly describe the facility's operating status at the time of the occurrence. As the name implies, this document is to be in narrative form and should be concise (one to five paragraphs is adequate). The narrative should answer the following questions:

1. What event has occurred?
2. How was it discovered?
3. What time did it occur?

4. Where did it occur?
5. What happened?
6. What response has been taken?
7. What damage has been reported?
8. What will happen in the immediate future?
9. Will the event move geographically?

. The narrative should be written in such a manner that it will get the exercise participants "in the mood" to set the stage for the actual exercise.

Master Sequence of Events List (MSEL)

The second portion of the simulation package is a listing of all of the events that are to occur during the simulated incident. This list of detailed events should begin where the scenario narrative ends and be designed to trigger the actions that are necessary to meet the exercise's objectives. The first step in the development of the MSEL is a review of those objectives. For each of the objectives, determine an event that will cause the participants to perform those actions necessary to meet the objectives. The end result should be a step-by-step list of logical events.

The second step is the completion of an MSEL form, such as the one shown in figure 14–1. To use this form, record the identified events in the "Event" column on the form. Next, list the expected actions for each event in the right-hand column (these are the actions defined during the refining of objectives process). There are two columns provided on the form for recording time: "Actual" and "exercise". Actual time refers to normal clock time. Exercise time refers to the time elapsed from the start of the exercise and is important for major events since it allows better control than actual time. (Exercises do not always start at the planned time and often deviate from the schedule).

The MSEL form also provides a means of referencing the written problems and messages. The column marked "Cue #" is for recording the appropriate number assigned to the problem/message used to initiate each event. The MSEL should also include control points. Control points are those actions that must be taken if the exercise is to proceed as planned. The control points need to be clearly identified (by stating that the event is a control point) so that they can be monitored during the exercise. If the expected actions do not occur at a control point, the exercise controller will have to take steps to get the exercise back on course.

Master Sequence of Events List				
Exercise Name: _____			Date: _____	
Time		Event	Cue #	Expected Action
Actual	Exercise			

Figure 14–1: Sample master sequence of events list

Problems and Messages

The purpose of planned, written problems and messages is to provide a means of introducing the listed events to the participants. In an actual emergency, response personnel will be reacting to events or changing conditions that they can observe or are made aware of through other personnel, but during an exercise these conditions can only be simulated through problems and messages.

Problems.   A problem is a narrative description of events or an action by an exercise simulator, which require participants to take actions. In a real occurrence, such problems would be observable or measurable, but because many events/situations cannot be physically produced for just an exercise, the problem narrative is used during exercises. During orientation and basic tabletop exercises, the problem statement is very simple and general. Since the purpose of the problem is to initiate participants action, it can be stated as simply as "What do we do if there is a fire?" In advanced functional drills and full-scale exercises where it is necessary to trigger specific actions in order to ensure that all exercise objectives are met, more detailed and specific problems are necessary.

The "Exercise Problem/Event Simulation" form in figure 14–2 is intended for use by controllers, simulators, and participants of the exercise. To utilize this form, refer to the following instructions:

Problem No. P - _____

Exercise Problem/Event Simulation

Time : _____

Description of problem/event: _____
_____
_____

Method of transmittal/simulation: _____
_____
_____

Controller/simulator comments: _____
_____
_____

Figure 14–2: Sample exercise problem/event simulation form

1. Each problem should be numbered in order of occurrence to help in organization. This number is to be referenced on the MSEL in the column entitled "Cue #." Problem numbers have been given the letter prefix "P" to differentiate them from messages.

2. The time portion of the form would be used to indicate at what point the problem is to be used in the exercise. Normally, exercise (elapsed) time is shown.

3. On the lines so indicated, describe the problem or event in detail.

4. For "Method of transmittal/simulation," describe, to simulators or to the controller, how the problem is to be introduced to participants and what method(s) would be used to simulate the event in order to maximize realism. This section should reference simulation aides/materials that would be used.

5. The bottom portion of the form is provided for the controller/simulator, to record any comments concerning the problem for later use in evaluating the exercise.

Messages. Whereas problems are used to simulate observable events and situations, messages simulate communications. Messages can be used in any type of exercise but are seldom used in orientation and basic tabletop exercises. They are especially important during functional drills and full-scale exercises. Messages are written to answer the basic questions: Who sends what? To whom? How? An exercise message form should be used in the development, control, and use of messages. As was discussed in the section on the MSEL, messages should be referenced to the event to which it corresponds. The "Exercise Message" form in figure 14–3 provides a space for assigning a number that is to be referenced in the "Cue #" column on the MSEL. This number has been given the letter prefix "M" to distinguish messages from problems "P".

The message form also provides space for indicating to whom the message is addressed ("To"), from whom the message was generated ("From"), and the means of transmittal ("Via"). Details on the method of simulation, if any, would be written in the "Special Instructions" section. For example, messages can be called in via radio or telephone by a simulator or simply handed to a participant by a controller. The space designated for "Text" is to be used for the message itself. The language in the text should clearly communicate the required information. To ensure a degree of realism, the language used should be the same as would be expected during a real emergency. Space is provided on the form for recording the time at which the message is scheduled to be delivered. As is the case with problems, exercise time should be used rather than actual (clock) time since delays in the exercise would make using actual time confusing. Space is provided at the bottom of the form for the simulator or controller to make comments concerning the message and/or the participants actions, for use during the exercise evaluation.

```
                        ┌─────────────────────────────┐
                        │   Message No.  M -           │
                        │                              │
                        └─────────────────────────────┘

                              Exercise Message

                        ┌─────────────────────────────┐
                        │   Time :                     │
                        │                              │
                        └─────────────────────────────┘

   ┌──────────────────────────────────────────────────────────┐
   │   To: _____      From: _____          │
   │   Via: Verbal _____ Telephone_____ Radio _____ Written ____│
   └──────────────────────────────────────────────────────────┘

     Text: _____

     _____

     _____

     _____

     _____


   Simulator's Instructions and Notes:

   Message to be transmitted:  Simulated _____ Actual _____
   Source agency/person:       Simulated _____ Actual _____

   Special Instructions:

     _____

     _____

   Controller/Simulator comments:      _____

     _____
```

Figure 14–3: Sample exercise message form

LOGISTICS AND SUPPORT

Every exercise requires certain logistics and support, the degree of which will vary from exercise to exercise. By logistics it is meant that a certain amount of supplies, facilities, displays, and equipment will be needed to enable the exercise to be conducted properly.

Exercise Area

Regardless of the type, every exercise must have an area or areas where activities will take place. The basic rule to follow is that the exercise should occur in the same place that actual, corresponding emergency operations would occur.

Support Facilities

In addition to the operations area, several support facilities may be necessary, depending on the type and scope of the exercise.

A **simulation room**, where exercise personnel can send, receive, and track messages and other communications, will usually be required in exercises where the interaction of various supporting organizations are to be simulated. For example, in an exercise involving the operation of the EOC, a simulation room staffed by exercise personnel representing support agencies such as the fire department and fire response team, would be used to call in or send in simulated messages to the EOC.

A **message center** is a clearinghouse for exercise messages that come from controllers/simulators to participants. It serves to distribute messages to the appropriate organizations and/or participants. The message center need not consist of more than a desk, manned by a single person, located in the simulation room or operations area. It is not normally needed in any but the most complex exercises, since messages can just as easily be handed directly to a participant by the simulators or controllers. The message center sometimes is used to simulate a communications center, but in an austere exercise, unless one of the objectives is to test message referral, the message center may not be necessary.

A **control center** is another support facility that is probably necessary only in the largest of exercises. Exercises can usually function quite successfully without a special area designated for controller activities. The controller can usually operate in the simulation room or operations area where he or she can directly supervise the progress of the exercise.

Communication Equipment

The communications equipment used in an exercise should be the same as will be available in an emergency. However, there are certain other considerations that must be made when planning for communications in an exercise. To begin with, it must be realized that during an exercise, time requirements are greatly compressed. This is due, in part, to the limited time available for conducting the exercise. Also, many activities that in a real situation would take hours to complete are simulated and require only minutes in the exercise. The result is that the flow of messages may overwhelm the communications system. Therefore extra radios and telephones may be required.

Many exercises will be conducted while normal business activities continue, and some communications equipment, therefore, may not be available to the exercise participants. In an actual emergency, priority would be given to response-related activities, but this is not always true for exercises. Additional equipment or the greater use of written (simulated) messages may be necessary. If simulation rooms and/or message centers are used in an exercise, extra communications will be required.

Special Considerations for Facilities

In regard to physical facilities for the exercise, there are other special considerations that require extra planning. The compressed time frame; the presence of extra personnel in the form of simulators, controllers, and so on, who would not normally be there; and the need, in many cases, to continue normal operations, impose special demands on existing facilities and resources. Items requiring special consideration include:

- sufficient workspace for extra personnel
- areas for visual aids (e.g., slides and videotapes)
- parking
- refreshments/food areas
- restrooms for participants and additional personnel
- name cards/IDs
- extra supplies (e.g., paper, pencils, clipboards) for the exercise staff

Displays and Materials

The preparation of special displays and materials is a necessary task if the exercise is to be conducted with the highest level of realism possible. While many documents and materials, such as evacuation route maps, are already part of the response procedures, many specially prepared materials can be used to enhance the realism of exercises.

Audiovisual presentations can greatly enhance the level of realism. Slides, photos, or videotapes of actual emergencies can be effective during some exercises to increase realism. Recorded messages can also help simulate TV/radio announcements for increased realism. The use of charts, maps, and blackboards can be an effective and relatively inexpensive way of presenting necessary information;

such displays are especially useful for tabletop exercises. The use of overhead projectors and transparencies can also be an extremely useful tool when conducting orientation and tabletop exercises; they allow for the simultaneous presentation of information to relatively large groups, are inexpensive, easily prepared, and can be easily adapted to changing exercise conditions.

Computers can be useful in a variety of ways. Because of their ability to store, display, and print large amounts of information, they are useful in planning and preparing for the exercise. Some computers have graphics capabilities that can be used in the preparation and presentation of graphs, charts, and diagrams. Because information in a computer can be quickly changed and documents quickly revised, printed, and displayed, it can be a useful tool for the controller. Time schedules can be quickly adjusted during the exercise to adjust the flow of messages as well as to revise and distribute schedules to help ensure that all personnel are aware of changes. It is sometimes necessary to change some messages or to generate new ones, and this can be done with comparative ease on a computer.

Special directories should sometimes be prepared, as necessitated by the more complex exercises. When off-site emergency organizations are involved, it is advisable to use a telephone number other than the normal emergency number, since using the fire department's emergency number may interfere with legitimate incoming emergency calls. This is especially true if extensive and continuous communications between participants will occur during the exercise.

A number of devices can be used for increasing the level of realism during functional drills and full-scale exercises. Some examples of simulation devices that can be used include, but are not limited to, the following:

- moulage (simulated injuries)
- smoke "bombs"/generators to simulate fire smoke or airborne chemical releases
- damaged equipment (e.g., a "junk" car used to simulate a road accident)
- inert substances (e.g., water) to simulate hazardous materials spills

EXERCISE ROLES

It is necessary that, in order for an exercise to be successfully conducted, exercise staff be appointed to fill three important positions: **controller**, **simulator**, and **evaluator**. In the more basic orientation and tabletop exercises, one person may be able to perform all three roles. As an exercise increases in scope and complexity, the number of persons necessary to conduct it must increase as well.

Controller's Role

The primary function of the controller is to ensure that the exercise is proceeding as planned. By monitoring the flow of messages and the decisions/actions of the participants, the controller makes sure that the exercise is following the outline defined by its scenario narrative and MSEL. It is his or her responsibility to take any necessary action to keep the exercise on track. The exercise director or a member of the exercise planning staff should be appointed as the lead exercise controller since he or she is familiar with the exercise's objectives, concepts, and scenario. The number of controllers is dependent on the size and complexity of the exercise. Specific duties of the controller include the following:

1. Monitoring the sequence of events to make certain the exercise is proceeding according to plan.
2. Maintaining order and professionalism throughout the exercise.
3. Acting as a simulator for unanticipated events by introducing spontaneous messages as needed.
4. Discarding messages in order to slow down the pace of the exercise.
5. Adding messages in order to increase the pace.
6. Checking actions and decisions of participants to make certain that the exercise is on course.

Controllers also have the responsibility for conducting debriefing sessions after the exercise has been completed. Controllers need to be trained and/or have some experience if they are to do an effective job. An inexperienced controller should receive training in the following:

1. Monitoring the sequence of events.
2. Monitoring message flow.
3. Controlling spontaneous inputs by simulators.
4. Coordinating information among simulators.
5. Responding to unplanned situations.
6. Monitoring the overall conduct of the exercise.

During the exercise the controller must have a copy of the MSEL, all messages, and any other simulation materials.

Simulator's Role

A simulator is responsible for the minute-to-minute flow of the exercise. The simulator's function is to create an artificial reality by portraying personnel from groups or organizations that are not involved in the exercise, but would be involved in an actual emergency. For example, an exercise involving only the EOC staff might employ simulators to act the parts of the response team and/or off-site organizations. Because the role of the simulator is to portray a member of a group or organization, he or she must be familiar with that group's responsibilities, tasks, and capabilities. The duties of simulators include the following:

1. Simulating all action taken by the group or organization, using written messages.

2. Sending messages representing the group's expected responses and reports, according to the MSEL time schedule.

3. Responding to unanticipated actions of exercise participants by issuing spontaneous messages.

4. Informing the controller of any deviation from the scenario.

It is important to realize that simulators will not just be issuing messages to the exercise participants. They must also be prepared to receive information from participants and simulate the appropriate follow-up actions. Simulators need to be supplied with the appropriate planned messages and a copy of MSEL.

Training of simulators is important. Personnel must be made fully aware of the sequence of events and familiar with the messages to be read. Personnel must be capable of knowing when to stick with the planned materials and when to create their own messages. Training should familiarize the simulators with the following:

1. Objectives of the exercise.

2. Message forms and flows.

3. Content of all exercise messages.

4. Development of spontaneous messages.

5. Coordination with other simulators.

6. Accuracy, timeliness, and realism of responses.

7. Interaction with the controller.

The number of simulators required depends on the size, type, and complexity of the exercise. Using personnel who have a working knowledge of several organizations allows for fewer simulators. If using a single person (or only a few people) to simulate all or many groups would detract from the level of realism, then additional simulators would be necessary.

Evaluator's Role

An evaluator is a necessary element in an exercise if the primary function of evaluating response readiness is to be achieved. Evaluators observe the exercise and afterwards report what went wrong and what went right. Where possible, evaluators should be obtained from organizations not directly involved with implementing the response plan. However, evaluators must have an understanding of emergency response concepts, as well as be familiar with the facility's plan. A possible source of evaluators would be corporate personnel or personnel from neighboring facilities not involved in the exercise.

Since evaluators must be drawn from groups that may have an incomplete understanding of the plan and basic exercise concepts, training should include:

1. Explanation of the objectives of the exercise.
2. Familiarization with participating organizations.
3. Message flow procedures.
4. Specific decisions or actions to observe.
5. Debriefing procedures.
6. Procedures in preparing an evaluation report.
7. Familiarization with the plant's emergency plan.
8. Familiarization with the plant's emergency organization.
9. Review of all exercise messages and the MSEL.

EXERCISE OUTLINE

The final step in the planning process is the preparation of an exercise summary outline. This document should be a complete summary of all the details relating to the exercise. Its purpose is to serve as a control document, to be used in presenting the exercise to management, and as a basis for developing the necessary training sessions for exercise participants and staff. While its format and content may vary from exercise to exercise, the outline should have an introductory section containing the statement of purpose, objectives participants, duties and time schedules, costs, and necessary staff assignments. This section should include a brief description of the exercise, indicating what actions are to be taken by participants, what equipment is needed, what actions and events are to be simulated, and who will simulate them. Information on precautions to be taken and any other administrative details that are necessary to carry out the exercise shall also be included. The exercise outline should also include all exercise materials and forms such as the MSEL, messages, problems, evaluator forms, and so on.

15

Conducting and Evaluating the Exercise

The culmination of all the effort placed into planning and developing the exercise is its actual conduct. The purpose of this section is to provide guidance and techniques for beginning, sustaining, and ending the exercise. It will attempt to define some common problems that will be faced and solutions to those problems.

CONDUCTING ORIENTATION EXERCISES

Because of their limited scope, the planning and preparation for orientation exercises are minimal, and the techniques and efforts required to conduct this type of exercise are also simple. Orientation exercises are scheduled events, whereby the time and place for the exercise is announced to all participants in advance, and both the beginning time and the ending time are set.

The orientation should begin with a brief introduction by the exercise coordinator. This introduction should include an explanation to all participants that the primary purpose of the exercise is to test their understanding of the emergency plan's provisions. The plan elements to be discussed during the exercise should be clearly identified. The exercise controller should then introduce a scenario narrative in order to initiate a discussion of how the emergency plan provides for response to the problem. Because the purpose of the exercise is to measure the participants knowledge and understanding of the emergency plans procedures, it is best to guide the discussion by using a step-by-step approach. A simple technique to use is to ask questions such as "What is done first?" and "What comes next?"

The orientation exercises are not intended to be lectures. The exercise controller should limit his or her participation in the discussion to controlling the flow and direction. This can be done by redirecting discussions, asking questions, and involving all participants in the discussion. The exercise controller should not be concerned with correcting mistakes made by any of the participants during the exercise unless the mistake is directing the discussion away from the objectives of the exercise. Mistakes and misconceptions of the emergency plan can be dealt with during the postexercise debriefing and during both the evaluation and follow-up phases that will be discussed later.

Orientation exercises will end when the participants' discussions have succeeded in meeting the objectives of the exercise or when time scheduled for the exercise runs out. Since orientation exercises are normally scheduled to end at a set time, the exercise controller must attempt to provide sufficient time to meet all objectives and still allow time for debriefing. If all objectives cannot be met, the exercise controller can elect to continue beyond the scheduled time (if participants' schedules allow), reschedule further discussion, or end the exercise at that point.

CONDUCTING TABLETOP EXERCISES

The process of conducting a tabletop exercise is similar to that of an orientation exercise. However, tabletop exercises can vary greatly in complexity, scope, and level of realism. In reality there are two levels of tabletop exercises: basic and advanced. Basic tabletop exercises are, as in the case of orientation exercises, concerned with the resolution of a basic problem through group discussion. Although somewhat more time should be allowed for tabletop exercises than for orientation exercises, they are conducted in much the same way as orientation exercises, beginning with an introductory briefing covering the purpose, scope, and administrative ground rules, followed by the introduction of the scenario narrative by the exercise controller. The scenario is the starting point for a discussion of the plan's provisions and procedures, which should include details on specific locations, level of severity, and other related issues. The exercise controller must control the flow and direction of the discussion to ensure that the exercise objectives are met. This is done by following the same general rules and techniques used in orientation exercises. A basic tabletop exercise will end when all objectives have been met or when the scheduled time period is over. The exercise controller is faced with the same decision of continuing on until complete, scheduling a later continuation session, or simply ending the exercise if all objectives cannot be met in the time allowed.

Advanced tabletop exercises employ the same techniques as do basic tabletop exercises; however, advanced tabletop exercises add the extra element of introducing a series of secondary problems to the basic problem presented in the scenario narrative. The advanced tabletop exercise begins with a briefing and the introduction of the scenario narrative; however, as the discussion continues, the exercise controller will introduce a series of related problems or events that require participants to discuss solutions to each problem. Some important characteristics of the tabletop exercise are as follow:

1. The advanced tabletop exercises would require the development and use of an MSEL.
2. Events are introduced to the participants through messages.
3. Messages may be introduced to all the participants for free and open discussion or be directed at particular individuals. For messages directed at a specific person, that person would outline his or her response or solution, which would trigger discussion by the other participants.
4. It is up to the exercise controller to monitor the direction and flow of the discussion so that all messages can be introduced within the scheduled time frame. Also, the order that they are introduced may have to be altered to fit into the context of the conversation.

The same general comments that were made concerning the ending of orientation and basic tabletop exercises also apply to advanced tabletop exercises as well. However, it is important that the advanced tabletop exercise be completed rather than suspended for lack of time. The exercise is complete when all problems have been resolved to the satisfaction of the exercise controller.

Advanced tabletop exercises require the preparation of maps, displays, overhead transparencies, photos, and so forth, to aid in conducting the exercise. Since the exercise is conducted in a classroom-type environment and not "in the field," display materials are of great value.

CONDUCTING FUNCTIONAL DRILLS AND FULL-SCALE EXERCISES

The methods used in conducting both functional drills and full-scale exercises are essentially the same, differing only in scope and complexity. Both types involve the highest level of realism, and they both involve the actual performance of many response tasks. They are "field" exercises, meaning that the exercises are to take place in the same types of location as would real emergencies. Both differ considerably from the manner of conducting orientation and tabletop exercises.

An important difference between orientation/tabletop exercises and functional drills/full-scale exercises is that while the former have an announced starting time and date, it is sometimes desirable not to inform the participants of the exact time frame for functional drills/full-scale exercises. This "no-notice" type of exercise is desirable when an objective of the exercise is to test the warning and notification procedures. Without this element of surprise, it is not always possible to know whether participants are responding to the unexpected notification or simply showing up as scheduled.

It is important that even in no-notice exercises, participants be briefed on the objectives and details concerning the exercise before it begins. The success of the exercise depends on the participants having a clear understanding of what is expected of them. This can be done in a exercise briefing, which can be conducted sometime during the week before the exercise. The exercise briefing should include information on: (1) how long the exercise will last, (2) who will be involved, (3) safety precautions to be followed, and (4) reporting/recording procedures. Details on the exercise scenario should not be given to the participants since this would allow for extra preparation on their part. Administrative details such as the location of restrooms, lunch periods, and so on, should be provided in a written handout.

The method of beginning a functional drill/full-scale exercise may vary depending upon the exercises objectives. However, since the majority of functional drills and full-scale exercises have the testing of notification/alarm systems as an objective, they normally start when the exercise controller introduces the initial message that simulates an initial discovery of the emergency situation. Unlike orientation and tabletop exercises, where participants gather at a predesignated location at a scheduled time prior to the exercise, participants in functional drills/full-scale exercises would not respond until the first message is released. In other words, they would continue their normal activities until they are notified of the start of the exercise through the same means as they would be in an actual emergency. For example, fire response personnel would not respond until the fire alarm is sounded. To avoid any confusion or panic, all exercise messages—especially the initial report—should be preceded and ended with the statement, "This is an exercise . . . ." If alarm systems are used, the PA system should be used to announce a drill just before and immediately following an alarm.

Depending on the exercise's objectives and scope, there are several different ways of introducing the initial message. If the emergency condition would be identified though automatic detection devices or the monitoring of instruments, the exercise controller should manually trip the alarm, if possible, or advise the person normally assigned to monitor the instruments of the simulated emergency situation. If the exercise does not involve the initial notification/response activities that would occur in an actual emergency, the exercise controller would use the scenario narrative to inform participants of the "current situation" within the exercise, followed by the initial message/problem.

To obtain the highest levels of realism, the exercise participants should be required to actually perform as many response tasks as practical. For example, it may not be practical to perform firefighting tasks involving the use of fire hoses since the water could cause damage. However, participants should be required to hook up and lay out the hoses and perform all other tasks short of actually turning on the water.

Once the initial message begins the exercise, it is the responsibility of the exercise controller to insure that all further messages are introduced on schedule according to the MSEL. It is the controller's responsibility to ensure that the exercise flows at an even rate and stays on track. Another problem faced by the exercise controller during functional drills/full-scale exercises is that it is sometimes necessary to stop the exercise momentarily in order to advance the simulated time "hours" or "days" ahead. Remember, activities that would take long periods of time to complete in a real emergency must be simulated within the compressed time frame of the exercise. For example, where it would normally take several hours, or longer, to bring a major structural fire under control, in an exercise this would be reduced to several minutes. To accomplish this, the controller should halt the exercise after all initial response activities are complete and simply state to all participants that it is assumed to be several hours later and the fire is out. The exercise activities would then continue from that point.

Functional drills/full-scale exercises normally end when all exercise objectives are met (signaled by the completion of the expected actions on the MSEL) or when scheduled time expires. Because of the disruptions to normal operations and problems with scheduling, it is normally undesirable and impractical to continue an exercise beyond the scheduled time or to reschedule a continuation. It is therefore imperative that the exercise controller keep the exercise on schedule or make any necessary adjustments early in the exercise so that essential objectives can be met.

Since the exercise is intended to simulate emergency messages in a realistic manner, it would be very easy for participants to mistake a notice concerning a real emergency for an exercise-related message. It is best to use a code word or phrase, agreed upon in advance, to signal the immediate end to the exercise and to make clear that all future messages should be considered real by all participants.

EVALUATION

Adequate provisions for evaluating the performance of tasks and duties during the exercise is necessary. The primary goals of the evaluation are as follow:

1. To identify deficiencies in the emergency plan/procedures.
2. To identify training and staffing needs.
3. To determine adequacy of equipment and resources.
4. To determine if the exercise achieved its stated objectives.

What to Evaluate

Determining what is to be evaluated requires a degree of planning and preparation, including a review of the documents and information used to prepare for the exercise. The first step in determining what is to be evaluated, is to review the specific objectives of the exercise. When the objectives are first developed, consideration should be given to the evaluation. If the objectives involve resource allocation or communications, be prepared to evaluate that particular item. Criteria for the evaluation of each specific objective should be considered during development of the exercise. If it cannot be measured or evaluated, it should not be considered an objective.

The next step is to review the key events from the MSEL for each of the objectives. The person or group responsible for performing the expected actions needs to be identified. This identifies who is to be evaluated doing what. The locations and time in which each action is to be performed also need to be identified so that an adequate number of evaluators can be determined. To be effective, evaluators must be stationed near enough to the action to clearly observe the participants. To aid in the organization of this information, an evaluator's exercise summary form, such as the one in figure 15–1, should be used during the exercise. The purpose of the form is to:

1. Outline the exercise objectives.
2. Define the expected action of participants so that the evaluator can compare this to what actually occurs.

Evaluator's Exercise Summary				
Exercise Objectives	Expected Action/Decision	Participants	Location	Time

Figure 15–1: Sample evaluator's summary form

3. Record the difference between expected and actual actions.

How to Evaluate the Exercise

The exercise is to be evaluated in three stages: (1) evaluators' reviews, (2) participants' debriefing, and (3) exercise critique.

Evaluators' reviews. The primary evaluation of an exercise is to be based on the observations of evaluators, controllers, and simulators. These personnel are in a position to observe and record the responses of participants, but because they are not participants themselves, they will be more unbiased in their evaluations. By observing the actions/decisions taken by the participants during the course of the exercise and comparing these to the expected action, evaluations can be made, based on the following:

1. Adequacy of procedures will be made apparent by the effectiveness of the response.

2. Staff inadequacies will be apparent when participants cannot complete tasks or have difficulty accomplishing tasks on a timely basis, or if tasks are accomplished in a hurried manner.

3. Equipment needs will also be made apparent by analyzing problems experienced in accomplishing tasks using the equipment on hand.

4. Training inadequacies will come to light through analysis of mistakes in judgment and poor execution.

Immediately following the exercise, the evaluator's comments should be reviewed by the exercise staff by comparing them to the evaluator's exercise summary and to the MSEL. Any and all areas where there are discrepancies between the expected actions from the MSEL and the actual responses of exercise participants, should be looked at to determine if any changes in the plans procedures, training programs, and/or equipment are appropriate. Information from the forms and the initial staff review is to be included in an exercise evaluation report that is to be prepared by the emergency coordinator.

Participants' debriefing. Many emergency plan deficiencies can be identified through an immediate review of the exercise by the participants themselves, since evaluators will not be able to catch every possible problem that occurs during the course of the exercise. In exercises involving many organizations, individual debriefing sessions should be conducted for each organization under the direction of a member of the exercise staff. When small groups are involved, debriefings should be done verbally with each participant being asked for comments in turn. In large groups, written comments should be requested. A common problem in meetings of this sort will be getting the participants to limit their comments to an evaluation of the emergency plan and response activities. The tendency will be to comment on the conducting of the drill itself, which can be avoided by telling participants that a critique of the exercise administrator will be conducted following the debriefing.

Exercise critique. This evaluation is different in that its objective is not to evaluate the emergency plan and response activities, but is intended to evaluate the administration of the exercise itself.

Exercise Critique

Please take a few minutes to complete this form. Your opinions and suggestions will help us prepare better exercises in the future.

                                                 Circle one

1. Did you understand the purpose and objectives of the exercise?   Yes No

2. Do you feel the purpose and objectives were met?   Yes No

3. Was the scenario narrative understandable?   Yes No

4. Do you feel the scenario was realistic?   Yes No

5. Do you feel the pace of the exercise was: too slow ? _____
                                   just right? _____
                                   too fast ? _____

6. Please rate the exercise overall, using the scale below:

1	2	3	4	5	6	7	8	9	10
Very									Very
Poor									Good

7. How would you compare this exercise to previous exercises?

1	2	3	4	5	6	7	8	9	10
Very									Very
Poor									Good

8. Did this exercise effectively simulate the emergency environment and test your emergency responsibilities? Yes _____   No _____

9. Please list any problems and/or your suggestions for improving future exercises:

_____

_____

_____

Figure 15–2: Sample exercise critique form

Exercise critique forms, such as the one in figure 15–2, should be distributed to all participants and exercise staff with instructions that it be completed immediately.

Final Exercise Evaluation Report

A final evaluation report, complete with recommended corrective measures and schedule for corrective actions, should be prepared by the exercise director. As soon as practical, the comments from the evaluations should be compiled and reviewed by appropriate personnel from the participating groups. A meeting of the exercise planning staff should be scheduled within a few days of the exercise to discuss these comments and to determine appropriate corrective measures. The exercise director should prepare and submit to the general manager (and heads of all affected off-site organizations) a final report summarizing the exercise results, including the following:

1. A summary of the exercise, including a review of the purpose, objectives, and scenario used.

2. A summary of major discrepancies/deficiencies.

3. Recommendations and corrective measures.

4. A schedule for the completion of these corrective measures.

FOLLOW-UP

Publishing a report on the faults in the emergency plan, training deficiencies and equipment needs and suggestions for correcting them is not sufficient to ensure that the exercise will result in positive action. The emergency coordinator is responsible for monitoring the progress of the corrective measures. The emergency coordinator is responsible for publishing and distributing, as appropriate, all changes in the emergency plan that result from the exercise. The emergency coordinator, based on the yearly exercise schedule, and with consideration to the cycle of exercises, should begin planning for the next appropriate exercise. Based on the results of the exercise critique, the emergency coordinator is to make necessary changes in exercise procedures in order to improve all future exercises. Upon completion of all corrective measures at the plant, the emergency coordinator should file a final report with the plant manager.

16

Community Education and Awareness

Effective industrial emergency response programs can be considered inadequate by the public if the community is unaware or misinformed about it. When your plant sponsors a community education and awareness program, the benefits are twofold: (1) The public will learn of your facility's efforts to ensure its safety, and (2) the information offered in the program might help to protect the public during emergencies.

COMMUNICATING RISK

When presenting or communicating risk to the public, it is important to present the concept in relative degrees of risk, rather than as an absolute, in order to place an issue in a broader context. There are many methods of doing this, some better than others and all with drawbacks. You should realize that you will not always be able to change people's perspectives, but using the following approaches may help.

One approach is to present your facility's risk(s) in comparison with other risks. For example, comparing the pollution emissions of your plant to state and federal standards might be useful. There are always some people who feel government regulators are in cahoots with industry, but for the majority of concerned citizens, knowing that a nearby facility meets or exceeds government standards is comforting.

Comparing the risk to other commonly accepted risks is also useful. For example, a renowned researcher once stated that the risk of a serious accident at the Shoreham Nuclear Power Plant was less than the risk of a devastating earthquake in New York City. This comparison helped a number of people to understand the value of the plant's built-in safety features. Unfortunately for the utility, the comparison did not work. There remained some strong opponents to the plant that dismissed this comparison saying that they did not have any control over nature but they could exert control over the licensing of the plant.

It is also useful to present health risks in the context of options and trade-offs, rather than viewing the risks associated with a single action. For example, a pharmaceutical plant might manufacture an important drug that benefits thousands of people, yet production produces only minor risks. When using this type of argument in combination with a description of plant safety and risk-reduction programs, most members of the public tend to support you.

Regardless of how well-thought-out a risk communication technique may be, there must be a fostering of trust between the facility representative and the public. If the public does not trust the bearer of information, they will not believe the information that he or she is disseminating.

IDENTIFYING COMMUNITY CONCERNS

Facilities often believe that there are no problems with their neighbors, yet when a problem occurs, many dormant and ignored concerns surface. In developing a community education strategy, it is important to evaluate community concern relative to your facility, if any, and to gauge the potential intensity of these concerns. The best way to identify these concerns and gauge their intensity is to talk to state and local officials (bureaucrats and elected leaders) and, more important, to area residents.

Also look at the recent history of the plant as reported in local tabloids for signals of what the public's perception of the plant might be. Have there been many plant layoffs in recent times? Has the facility been a principal party in many environmental emergencies and/or ongoing environmental problems, such as pollution? Have there been many identifiable benefits to having the plant in the community (again, concentrate on the public's perception rather than on your grasp of the facts)? When reviewing the local newspapers, look beyond stories about just your facility. How are environmental issues handled in general? Is there an increased awareness and concern by the local community as a result of stories of other facilities' programs?

In the preliminary investigative phase, address the following questions:

1. Does the public perceive an environmental, health, or safety problem with industry in general? With your facility specifically?
2. Does it perceive a lack of interest on the part of the government for the welfare of the people living near the plant?
3. Does it perceive a lack of interest on the part of the company for the welfare of the people living near the plant?
4. Is there any public resentment due to nonenvironmental issues (e.g., recent layoffs)?
5. Is the company an important part of the economic well-being of the community?

By performing this type of investigation, you will be able to identify and rank the public's concerns and develop objectives and techniques for dealing with the issues.

OBJECTIVES AND TECHNIQUES

While your community education process may not be able to alleviate all community concerns, problem issues will not disappear if they are ignored. Also, an emergency education program designed for the public does not necessarily have to solve community relations problems but rather can simply convey information about what to do during an emergency. For every objective in a typical emergency education program there are specific and reasonable techniques to use, many of which are enumerated in the following:

Objective: To provide facility neighbors with accurate and complete information about environmental, health, and safety issues; plant processes; and other facility programs.

Techniques:
1. Establish and maintain a library or information repository relative to the subject areas of concern and provide public access to it.
2. Conduct tours of your facility and seminars about your facility for area residents.
3. Issue fact sheets or brochures containing relevant information about these programs.

Objective: To provide area residents with information about ongoing safety and emergency response plans and programs, and, specifically, information about what to do in case of a facility emergency.

Techniques:
1. Issue brochures containing a summary of plant safety and emergency response plans and information about what to do in the event of emergency.
2. Issue concise information about what to do in the event of emergency. This information can be printed in or on
   - emergency public information brochures
   - telephone books
   - shopping bags
   - refrigerator magnets
   - stick-on labels
   - calendars
   - Other handy items typically found in a household.
3. Conduct briefings at local community groups including schools and fraternal or business groups (e.g., Rotary, Elks, Kiwanis)

Objective: To establish and maintain contact with state and local officials concerning the aforementioned items.

Techniques:
1. Conduct briefings concerning program developments.
2. Conduct regularly scheduled telephone briefings of public officials to assure that they are apprised of program activities.
3. Include public officials on the mailing list for the distribution of information outlined above.

Objective: To ensure that media sources are provided the same information in a timely manner.

Techniques:
1. Issue press releases to coincide with major program activities.
2. Conduct media briefings concerning program developments.
3. Conduct regularly scheduled telephone briefings of media reporters to assure that they are apprised of program activities.
4. Include the media on the mailing list for the distribution of information outlined above.

Objective: To periodically reassess the community's concerns.

Techniques: 1. Periodically perform responsiveness surveys of public officials and the general public.

2. Conduct content analysis of recent articles of media reports about your facility and environmental issues.

17

Dealing with the Media During Emergencies

The purpose of this chapter is to impress upon the reader the need for effective media relations during emergencies. This chapter provides information concerning: (1) the need to develop an emergency public relations plan, (2) the characteristics of the media and how the media operates, and (3) communicating with the media, including how to conduct press conferences and interviews.

MEDIA CHARACTERISTICS

General Characteristics

The media is your liaison to the public during emergencies, and this is a fact that cannot be ignored. If you want to maintain good relations with the public, you must also develop and maintain good relations with the media. During an emergency, you must be prepared to properly present the facts to the media, and you must do it promptly. News is a competitive business, and if you do not cooperate by providing facts about the emergency before media deadlines, the media will go elsewhere for the story, oftentimes resulting in bad coverage for your firm.

You should realize that the media are basically predictable in the way they respond to emergency and disasters: their questions are predictable; their deadlines are predictable; their behavior is predictable. Because of this predictability, there is no reason not to be prepared to work with and respond to media demands.

Media relations do not have to be adversarial, and in fact should not be. Dealing with the media during emergencies is no different from any other emergency function. The better prepared you are to manage the function, the better the result will be. This is not to suggest that effective media management will turn bad news into good news, but it can prevent bad news from becoming worse. The public will judge a company's emergency response performance, for the most part, on the basis of how it is reported by the media, and therefore, knowing how to deal with the media in time of emergency is essential.

Print Media versus Electronic Media

The media do not comprise a homogeneous group and they are as diverse as any other profession. For example, not all attorneys practice the same law—some are criminal lawyers, some are tax lawyers, some are corporate lawyers. You should recognize the differences between the print and electronic media.

The print media is made up of newspapers, periodicals, and the wire services. They provide facts, analyses, and commentaries concerning the events covered. The stories build day after day and are not confined by time. By comarison, the electronic media is made up of radio and television. Television covers events with sight, sound, immediacy, and motion. Their stories are often compressed and short on detail. Television makes its impact by image, repetition, and timing and is aimed at a home audience. Radio is different in that it can reach audiences nearly everywhere they might be, not necessarily just at home. Radio also relies on immediacy and repetition to impact its audience. While many of its stories are also compressed, radio can afford to put more detail into its stories than can television. While deadlines differ among the different forms of print media, they are usually not as immediate as those of the electronic media. A newspaper reporter covering an event that occurred at 5:00 P.M. may not have a deadline until midnight, while the television reporter may need the story by 5:30 P.M.

Three basic characteristics to keep in mind are:

1. Television needs visually stimulating backdrops for its stories.
2. The electronic media have more pressing deadlines than do the print media.
3. The print media want more background information and details than do the electronic media.

It is best not to favor one form of media over another.

Local Media versus National Media

For most emergencies, you will deal with only the local media. Local and regional newspapers will send reporters to the scene; the wire services and local radio stations will probably phone for details; and some local television stations will send a reporter and camera crew to the scene. The national news media usually cover only extremely large-scale emergencies.

Both local and national media organizations have the same needs and should be treated equally. The tendency for most companies is to favor the national media, which is understandable since it must be difficult to turn down an interview with a nationally prominent newscaster. But this can be a costly mistake! The day after the emergency is over, the national media will be gone but the local media will still be there providing follow-up coverage. It is the follow-up coverage and analysis that will have the most impact on the public's and government's perception of the incident. It is therefore in your best interest not to anger the local media by ignoring their initial concerns.

PRE-EMERGENCY PUBLIC RELATIONS PLANNING

The key to good emergency media relations is planning and establishing a positive relationship with the media before an incident occurs. Develop an emergency media relations procedure as part of your overall plan, including the following elements:

1. Policy statement regarding media relations.
2. Statement of duties for a designated spokesman, media center coordinator, security personnel, and others likely to have public relations roles.
3. Guidelines for the "dos and don'ts" of dealing with the media.
4. Notification procedures.
5. Equipment and special features for the media, including a media center, safety equipment, communications facilities, interview and debriefing area, and certain supplies.
6. Briefing kits that include background information about the company and the plant, its operations, safety and environmental protection programs, key contacts, photographs and diagrams, etc.

With the completion of this procedure and subsequent staff training, your plant will be better prepared to deal with the media, but there still are other aspects to consider. You must now lay a foundation for a trusting relationship with the media, and this can best be accomplished through education. It is important that the media understand your firm, its products, its commitments and achievements, its processes and problems—prior to an emergency. The importance of this cannot be understated. Put yourself in the reporter's position, and think of how trustful you would be if the first time you met a company spokesperson was following a disaster that seriously affected the community. Would you not be somewhat wary in believing what he or she might be telling you? Contrast this to a situation in which you, as the reporter, had met plant personnel, had been brought on tours of the facility, and had gained at least a basic understanding of the plant's processes and safety components. You would then be in a better position to believe and understand what the company spokesperson was telling you and would have a better basis for asking informed follow-up questions. Meeting and educating members of the media prior to an emergency is therefore an important element of the media relations program. It is not that these efforts are an attempt to coopt the media, which is difficult if not impossible to do anyway, but rather are a basis for fair and accurate reporting if a crisis does occur.

Regularly scheduled meetings should be set up between company officials and local media. Senior management as well as technical staff should be involved in these meetings, not just the media relations coordinator. These meetings might include conducting tours and providing technical briefings on the processes used. Areas of special interest would be processes that deal with waste treatment, pollution control and the like. Be sure to highlight safety features, and, if possible, allow the media to observe emergency drills and exercises. The media relations coordinator should follow up on these briefings to ensure that the reporters understood the information and have an opportunity to clarify any information they might have misunderstood. There are examples when reporters misunderstood certain terms that were used during these types of briefings and went on to write very inaccurate and unfavorable stories.

Showcase company-sponsored events that reflect the relationship between the company and the public (e.g., benefits, scholarships, work-study programs). Be sure that you promptly follow up on any reasonable requests that might be made by reporters. A reporter should feel that he or she can count on your cooperation all the time, not just when it is convenient for the company.

COMMUNICATING WITH THE MEDIA

You should realize that today's news environment requires conflict in a story and that you must deal with it accordingly. The media will search out antagonists of your firm for an opinion, and if you do not respond accordingly, your side of the story will never get out. The media are not generally out to get you, but if they perceive a cover-up, they will pry into every aspect of your operation. Therefore, it is important to quickly provide the media with clear, concise, and accurate information.

Press Releases

Immediately following an emergency incident, or as soon as possible, the media should be notified. You will be judged to be forthcoming and credible if you talk to the media before they learn of the incident from other sources. One effective tool for providing information to the media is the press release. This can be distributed in person or by telefax, telegram, or mail. The fastest means of distribution is usually the most desirable.

A good emergency press release answers these basic questions concerning the event:

1. Who is involved?
2. What is taking place?
3. Where did it happen?
4. How did it happen?
5. Why did it happen?

The release should then go on to provide the details and background information. Also, be sure to include the name and address of your company, date/time of the press release, headline, and a contact person's name and telephone number if additional information is required.

Interviews

Think of the interview as an opportunity to get your side of the story across. Don't think that it is going to be a bad experience, or you may appear to be defensive. When answering questions, remember that it is okay not to know everything, and saying "I don't know" is appropriate as long as you commit to get the answer as soon as possible. On the other hand, answering "I don't know" to every questions can suggest evasiveness, so let only a knowledgeable representative of your plant be interviewed.

Remember the unique characteristics of the media. Quite often a reporter is short on time an space, and therefore, if you can anticipate the questions beforehand and prepare concise answers, you stand a better chance of being quoted directly (reporters love a good quote!). The following categories of questions are most likely to be included in an interview during an emergency.

Casualties.
1. Has anyone been injured? How many people? How many are hospitalized? Where?
2. Has anyone been killed? How many deaths? What are the dispositions for the dead?
3. Are the victims' names available? Do the victims include any of social or corporate prominence?
4. Are all possible casualties accounted for?

Property damage.
1. What are the estimates of on-site and off-site losses?
2. Describe the damage incurred. How vital/irreplaceable was the damaged property?
3. Is any other property threatened?
4. What is the extent of insurance protection?
5. Have there been previous emergencies in the area?

Causes.
1. What are the direct accounts of participants and witnesses?
2. How was the emergency discovered?
3. Who sounded the alarm? Who summoned aid?

Rescue/relief operations.
1. How many people are involved in the rescue/relief operations?
2. What rescue/relief equipment is being used?
3. What obstacle to the relief efforts exist?
4. Describe the care of victims.
5. Has the emergency's spread been prevented? How?
6. Has property been saved? How?
7. Describe any noteworthy acts of heroism.

Description of the crisis.

1. Describe any spread of the emergency.
2. Describe any secondary effects, such as blasts, explosions, crimes, and violence.
3. Describe any escape attempts.
4. What is the expected duration of the event?
5. Is a spill involved? To what extent?

Accompanying incidents.

1. What is the estimated number of spectators?
2. What unusual happenings have accompanied the emergency?
3. Describe the anxiety and stress of families and survivors?

Legal actions.

1. What has been determined through the coroners' reports?
2. Describe any ensuing inquests or police follow-up.
3. What insurance-company actions are being taken?
4. Has professional negligence been claimed? Are suits stemming from the incident?

If a reporter asks an unflattering question, you should modify or refocus it in a more positive vein. This is especially important when the reporter is not focusing on the key issues of a story. This is not to suggest that you are skilled enough to steer an interview, because most people are not, but rather you should be able to rephrase the questions to provide a clearer picture to the public.

If you anticipate being interviewed by a diverse group of reporters in a press briefing, you should develop a media briefing center that has audiovisual aids, telephones, adequate electricity, proper acoustics, tables, chairs, and other necessary equipment. Have an assistant attend the briefing with you to jot down questions that you have not answered or had difficulty answering. It is important that you have at your disposal other company experts to provide you with answers to questions you can't answer during the briefing, and immediately get back to the reporter(s) as soon after the briefing as possible.

After being asked a question, it is a good practice to repeat the question. This gives other reporters and listeners another opportunity to hear the question. It also allows listeners to hear your understanding of the question, and provides the opportunity to restate the question in terms that you believe are fair and pertinent to the matter. You must be careful not to restate a tough question in such a way as to avoid answering it, but it is appropriate to restate a question in such a way as to eliminate abrasive adjectives and tone from a reporter's original question. This is also a way to correct implied inaccuracies in a reporter's question in a manner that is not embarrassing to the reporter. Always begin your answer with "The question was . . .."

Reporters often ask a string of questions that are difficult to remember and extremely difficult to answer coherently. When asked a compound question, write down the key components of each question and restate the portion of the question that you intend to answer. Begin you answer with, "There were several questions asked. The first one was . . .."

Follow these basic rules of "dos and don'ts" during an interview:
During an emergency, **DO**:

1. Release only verified information.
2. Promptly alert press of relief and recovery operations.
3. Escort the press everywhere on the site.
4. Have a designated spokesman (and backup).
5. Keep accurate records of all media inquiries.
6. Identify and meet media deadlines.
7. Provide equal opportunities for print and electronic media.
8. Know what can and cannot be released (company policy).
9. Coordinate media relations functions with other emergency functions.

During an emergency, **DO NOT**:

1. Speculate on the causes of the emergency.
2. Speculate on the resumption of normal operations.
3. Speculate on the outside effects of the emergency.
4. Speculate on the dollar value of the emergency.
5. Interfere with the legitimate duties of the media.
6. Permit unauthorized spokespersons to comment to the media.
7. Attempt to cover up facts or mislead the press.
8. Place blame for the emergency.

18

Emergency Facilities and Equipment

Emergency plans must be based on realistic assessments of available resources, including facilities and equipment. No emergency plan should be based on what may or may not be available in the future. If additional resources are required, then provisions should be made to obtain them, but emergency plans should still be based on current resources and capabilities. Changes can then be made after additional resources are a reality.

One purpose of developing an emergency plan is to effectively utilize all available resources. Planners often overlook resources that are available but just not obvious. For example, while some facilities may not have a designated emergency control center, a meeting room can be converted utilizing available equipment to serve as an emergency control center in time of emergency. While your facility may not have all the spill control equipment required, a neighboring facility might make such equipment available during emergencies on a loan basis. It is difficult to come up with one list of required facilities, equipment, and supplies, because each type of plant has different requirements. An inventory of resources should be developed and included in the emergency plan. It is important that the inventory include the quantity, condition, and availability of all resources. Emergency acquisition procedures should be developed for resources that might be required during an emergency but not stocked as a matter of routine at the plant.

ON-SITE FACILITIES AND EQUIPMENT

This section identifies the types of on-site facilities and equipment that are generally necessary to support emergency operations.

Emergency Operations Center

The emergency operations center (EOC), sometimes referred to as an emergency control center (ECC), is a place from which the operations to mitigate an emergency are directed and coordinated. It is usually staffed by senior managers and coordinators of the emergency response team, and the room should be equipped with adequate communications equipment. A dedicated hotline or "call-out only" phone to an off-site response authority should be maintained and backed up by battery-operated radio equipment. Record keeping and incident-plotting status boards should also be included. Plant emergency plans, MSDS's (material safety data sheets), telephone directories, maps, community emergency plans, resource inventories, and other important reference information should also be maintained in the EOC. Office supplies such as pens and papers should be sufficiently stocked.

Media Center

A media center is necessary because it can be used to brief the media and to control the activities of media personnel who will inevitably arrive at the scene of significant accidents. The media center should be separate from the emergency control center so that operational personnel are not interrupted while performing their duties. The media center should contain desks and telephones for the media to use (coffee and doughnuts are optional). A plant security person and/ or a mid-level management spokesperson should be assigned to ensure that media personnel do not enter unauthorized areas where their safety or the safety of others may be jeopardized.

Accident Mitigation Equipment

Depending on the nature of emergencies that might be expected to occur, equipment and supplies to combat those emergencies should be available. Oftentimes, such equipment is required by federal or state regulation, (e.g., through the EPA or OSHA). Firefighting equipment and supplies such as fire extinguishers, fire hydrants, hoses, vehicles, rope, spill control equipment, absorbents, and so on, are necessary to control fires and hazardous materials spills.

Personal Protection Equipment

Personal protection equipment is required for those persons serving on the emergency brigade in accordance with OSHA regulations

[1910.156 (e) (1) (ii)]. Such equipment includes foot and leg protection; body protection; hand protection; and head, eye, and face protection. Regulation 1910.156 (f) also requires appropriate respiratory protection as well. A document published by the U.S. Department of Commerce, and available through the National Technical Information Service (#PB85-222230), *Personal Protective Equipment for Hazardous Materials Incidents: A Selection Guide,* provides a comprehensive guide to selecting the proper personal protection equipment for hazardous materials incidents.

Other Equipment

Other necessary emergency equipment might include medical (first-aid) supplies, accident detection equipment, meteorological equipment, and security/access control equipment.

Plant Warning and Communications Systems

An effective emergency response is contingent upon having an effective warning and communications system. Developing an effective system is contingent upon (1) selecting the proper equipment; (2) training workers to use the equipment and respond as required, and (3) inspecting and maintaining the equipment and procedures.

There are many different types of communications and warning equipment available on the market today. Choosing the proper equipment so that it can be put together in a comprehensive system is essential. Above all, this equipment must be capable of properly working in your plant's environment. Prior to purchasing or installing equipment, the following critical factors should be considered:

- plant processes
- machinery
- physical layout
- expected emergencies

Equipment should be installed that allows an employee to report an emergency to a central plant location. Typically, a fixed one-way alarm, such as a fire alarm mounted to a wall, is used as a reporting device. The advantage of this system is that it is usually already installed at most plants. It has a major disadvantage in that it does not provide any information about the incident. Two-way communications devices have the advantage of allowing information about the accident to be transmitted; however, the devices can be more expensive than one-way devices. Examples of two-way devices include conventional telephones, intercoms, cordless telephones, and radios.

Alarm equipment alerts employees to the danger and tells them, either through voice or prearranged signal, exactly what action to take. Examples of such equipment include audio equipment (horns, bells, chimes, whistles, sirens, public address loudspeakers) and visual equipment (flashing or rotating lights). A good alarm system should be loud enough so that it is capable of overriding ambient plant machinery noises, and of being clearly heard at the farthest point for which it is intended. It should be distinguishable from machinery and routine operational signals within the building and should be strategically placed so that it can be seen or heard at all locations within the plant. Finally, it should be flexible enough to alert employees to perform different types of response actions. For example, three different types of signals might be used to differentiate between plant shutdown, fire brigade mobilization, and plant evacuation.

There is no point in spending a lot of money on a plant signaling system if no one knows where it is located, how to operate it, what office numbers to call, or what the various signals mean. Therefore, it is imperative to train all employees on the mechanics and use of the alarm system. In addition, information that states what to do, whom to call, what information to report, and so on, should be posted at all critical points within the plant. Posters describing the meaning of various alarm signals should be prominently displayed throughout the plant.

No communications or warning system can work perfectly forever, and therefore, periodic testing of equipment is essential. Testing should be done on a periodic basis but when new process machines are added to an area or when new buildings are built, alarms should be tested again to ensure adequate coverage in these areas. Periodic retraining of employees is necessary as well.

Planning a Communications and Warning System

The following is an outline of procedures to follow in developing a plant's emergency communications system.

Step 1: Determine emergency communications requirements.
Key personnel in each department and from each primary off-site agency should be involved in determining which emergency re-

sponse groups need to communicate between which points and for what purpose. The basic source of information for determining emergency communications requirements is contained in the emergency functional assignments outlined in the emergency plans of the plant and of government agencies.

Step 2: Inventory existing communications resources.
Inventory all existing communications resources including telephones, hotlines, radios, warning systems, telegraphs, and public media. The inventory should include information regarding the characteristics, capabilities, limitations, and availability to meet emergency operational requirements.

Step 3: Match available communications inventory with requirements.
By correlating inventory data from step 2 to the requirements from step 1, available communications may be utilized to the maximum. This step will identify shortages and surpluses of communications equipment.

Step 4: Develop emergency communications and warning procedures.
Based on the results of steps 1 through 3, develop procedures to effectively utilize existing equipment and facilities. Make sure that procedures are written to utilize current equipment "as is," or without modification.

Step 5: Outline actions to correct deficiencies:
Develop a time-phased schedule for making necessary improvements to the existing system. A plan for acquiring necessary funding should be included.

Step 6: Update inventory and plans.
Modification of existing systems and additions of new systems and equipment will require changing the inventory and updating plans and procedures.

OFF-SITE FACILITIES AND EQUIPMENT

Most communities already have facilities and equipment necessary to respond to a wide range of emergencies. It is necessary to determine if these resources are adequate to handle emergencies at your facility. Each off-site emergency response agency should have a list of available resources. A combined community inventory is often found in the community emergency operations plan. When assessing these resources, the following criteria should be used:

- quantity
- availability
- limitations on use
- cost
- location
- authority to commit resources
- usefulness in combating plant hazards

Following your assessment of community resources, it is recommended that you make community officials aware of any deficiencies in their resource base so that corrective actions can be taken. When dealing with officials on this issue, remember that many communities do not have adequate budgets to purchase required resources and may look to industry for monetary support.

Off-site Warning Systems

The development and use of an off-site warning system is ultimately the responsibility of the local government. Many industrial facilities, however, have chosen to subsidize the development and maintenance of a public warning system.

The use of sirens is the most traditional method of warning the public of emergencies (however, not the only method). Sirens are in use around nuclear power plants, in areas prone to tornadoes, and in many areas that have active civil defense programs. If outdoor siren warning systems are used, it is essential to develop a method of following up the siren with instructional messages. Typically, a siren will be followed by emergency instructions on the radio and television. Some areas supplement outdoor warning systems with tone-activated radio systems that are also useful in conveying emergency instructions to the public. Very large, stationary public-address systems at the plant boundary have also been used to warn and notify the pubic of emergency conditions.

A guide from the Federal Emergency Management Agency, document CPG 1-17, entitled "Outdoor Warning System Guide," provides useful information concerning the principles of sound, various warning devices, planning a system, and testing.

19

Computers in Emergency Management

CONVENTIONAL SYSTEMS FOR MANAGING INDUSTRIAL EMERGENCIES

For essentially any business function today, a computer can be a useful tool in carrying out required responsibilities. There are several ways a computer can be used in an emergency preparedness program. Many useful programs are commercially available; however, their costs may be high, and their function may not match your exact needs. Some programs, such as database management systems, allow you to develop applications to suit your own needs, although a degree of competence with such programs is required. What follows are some applications that may be of use in your emergency preparedness program.

Hazard Assessment

Several companies currently offer microcomputer-based dispersion modeling systems that are useful in understanding the effects of potential releases. By providing theoretical data concerning meteorological conditions, time of event, and the type and quantity of chemical involved, a model of its dispersion (concentration, plume, etc.) can be generated. This is useful in developing plans for off-site emergency response; however, extreme caution should be used when this program is applied to actual emergencies. The software publisher's warnings must be read carefully to determine whether the program's effectiveness would be limited in a real emergency. Programs are also available to aid in risk assessment, such as systems that aid in the construction of fault and event trees that are used in hazard analysis studies.

Emergency Planning

Because emergency plans require frequent modification and updating, the use of a common word-processing program is beneficial and can save much time. Simple programs can also be developed so that individuals can obtain only those procedures that they are required to carry out during emergencies. These computerized emergency checklists can eliminate the reading of extraneous material during an emergency. One very useful type of automated plan is a hypertext-based emergency plan. Using this type of program allows the reader to dynamically read only those elements of the plan as is necessary. For example, assume a person is reading about his or her emergency duties and it mentions the term "cleanup techniques." If the person is unsure of all aspects of the techniques, he or she could simply "click" on the term and the hypertext program would automatically jump to the portion of the plan that discusses cleanup techniques. This avoids having to read through many needless sections of the plan, which saves valuable time during an emergency. Another example is that of an emergency coordinator who, during a lull in the emergency, is reviewing his or her emergency procedures. The procedures might state that he or she is responsible for briefing corporate officers as soon as possible. By clicking on the term "corporate officers," the hypertext program would jump to the part of the plan that lists the officers and their telephone numbers.

Event Management

There are a number of computer applications that can help in managing emergency incidents. These programs can help to assess the consequences of chemical spills, to keep track of emergency events, and to track resulting damages.

As mentioned, several companies currently offer microcomputer-based dispersion modeling systems. By providing data concerning meteorological conditions, time of event, and the type and quantity of chemical involved, a model of its dispersion (concentration, plume, etc.) will be generated. The models available today have limitations, however, and therefore should be used a tools only.

A database program can be used as an event-tracking log, which, following an incident, can be useful in reconstructing who did what, when, and why. This type of system can also be used to keep track of requests for information from off-site authorities or the media, which would be useful for ensuring that a response is provided to all requests. A database program can be used to track damage to plant facilities and equipment and the expenditure of supplies in combating the emergency. Such an application can later generate corporate damage assessment and insurance company reports.

Administration

Many commercially available programs allow a company to keep track of safety and environmental matters. These programs can often be applied to emergency preparedness related activities. Personnel training records can be maintained, program schedules kept, and inspection records maintained.

Resource Management

Database programs offer much flexibility in the design of resource management programs. Various types of equipment and supply information, such as quantity, condition, cost, and availability, can be kept. During an emergency, these resources can be monitored and a record of expenditures can be printed, thus ensuring that reordering will take place. While computers can be useful, they do not replace the work man does, but rather only increase productivity.

EXPERT SYSTEMS FOR MANAGING INDUSTRIAL EMERGENCIES

Imagine yourself in a situation where you are the weekend shift supervisor at an industrial facility. An employee accidentally crashes a forklift into several barrels of various chemicals. While you have received general training to handle certain spills, you cannot find specific information on how to handle this combined spill, what protective equipment should be utilized, and who should be contacted. Your company has an emergency plan, but it is a 1 1/2-inch thick volume. You need to take action now. The spilled material may be highly toxic, and its plume should reach the fence line shortly.

The problems outlined in this scenario are common to most emergencies—too little expertise to integrate and analyze all of the information available. One technology that may be of vital use in such a situation is a branch of artificial intelligence known as expert systems.

What Are Expert Systems

An expert system is a computer program with special problem-solving capabilities. The system relies on a database of knowledge about a very particular subject area, an understanding of the problems addressed in that area, and a skill at solving those problems. To a user, interaction with an expert system is like "talking" to a human expert through a computer terminal. The expert system asks the user several questions; explains why it has asked the questions (if necessary); offers conclusions, advice, and solutions; and explains the reasoning process it has used to get to the answers. In other words, an expert system emulates the human thought process and provides advice, as would a human expert.

To understand how expert systems imitate the human thought process, it is important to understand how a human expert may handle problems. First, a human expert has a formal body of knowledge and experience on a particular subject. Upon notification of a problem, a human expert will ask for specific relevant details, consider what he or she has learned, possibly ask for additional information, and then decide upon a course of action based on a set of rules the expert has developed over the years. This set of rules, referred to as heuristics, is usually in the form of "if-then" statements, although most people are not aware they think like this. For example, a simple "if-then" rule of thumb to which most people can relate is: If the traffic light is red, then I must stop the car. Over many years, experts develop thousands of rules that form the basis of their special bodies of knowledge.

Expert systems "think" in a similar manner. When the program is activated, it will ask the user a series of questions about the problem to be solved. The program will evaluate the information against the rules programmed into it. Expert systems often consider a number of competing theories and make tentative, weighed recommendations. Many expert systems also have the ability to describe their reasoning processes to the users. Does this suggest that expert system are without limitations or that they can replace human experts? Certainly not! Expert systems can solve difficult problems but only within a narrow domain. They also are incapable of reasoning from axioms or general theories and cannot learn (although recent technological advances are making this less true). The inability to learn means that they are limited to the specific facts and rules that were programmed into them. They also lack common sense and have a lot of difficulty when the problems are beyond the scope of the systems' designs.

On the other hand, expert systems do not jump to conclusions nor try to defend bad decisions in the face of contrary evidence. Their decisions are not biased (unless programmed to be biased). They do not have bad days. They are detail oriented and consider all possible solutions. They do not get tired. And the most advanced expert systems perform their specialized tasks at a human expert level. All things considered, an expert system can be a very valuable tool in solving difficult problems. As with any tool, however, inappropriate use may limit its effectiveness.

History of Expert System Development

Following World War II, groups of British and American scientists were working to develop the machine that would eventually come to be called a computer. The British wanted a machine that could follow programmed instructions and use logical operators such as "and," "or," and "not." Programs based on logical operators would be capable of handling statements in ordinary language. The American scientists knew that such a machine would be extremely expensive to build and, instead, chose to build a machine that would do only arithmetic calculations using numerical operators such as "+," "−," and ">." This decision resulted in the very fast calculating machines referred to now as computers.

In spite of the fact that the original computers were built as numerical processors, some scientists continued work on nonnumerical, symbolic processing systems. This work, combined with the work of psychologists concerned with human problem solving, evolved into a field of computer science known as artificial intelligence (AI). In the 1960s, a number of corporations believed that some of the work coming out of the AI research labs would be useful in business. Several systems were developed but proved to be too costly to develop, too slow, or unable to provide sufficiently practical results. The AI programs were just too complex to run on the computers in existence at the time. Improvements in microelectronics technology in the 1970s resulted in faster, more powerful, and relatively inexpensive computers that made the development of AI systems much more practical. One subset of AI that has matured in recent years is the expert systems technology now being used to develop very practical solutions to complex problems.

What Problems Can Be Solved with Expert Systems?

Not all problems are appropriate for applications of expert systems technology. But in general, problems arising from industrial emergencies contain many characteristics applicable to the technology of expert systems. Some of those characteristics are as follows:

*   occur often
*   are dynamic (variables change irregularly)
*   are qualitative rather than quantitative
*   require expert knowledge
*   require human experts that are not always available
*   require human experts that are too expensive to use on a regular basis
*   are complex and require the integration of large amounts of data
*   need a consistent response

Tools Currently Available to Emergency Responders

Before exploring the possible application of expert systems to emergency situations, it is important to understand what tools currently exist for emergency responders. Current tools fall into the following categories:

*   guides or handbooks
*   reference libraries
*   emergency response information centers
*   computerized databases

There are a number of guides and handbooks in use by emergency responders. The most common of these include the Department of Transportation's *Emergency Response Guidebook*; the Coast Guard's *Chemical Hazard Response Information System Reference Manual*; the National Fire Protection Association's *Fire Protection Guide on Hazardous Materials*; the Association of American Railroads' *Emergency Action Guides and Emergency Handling of Hazardous Materials in Surface Transportation*; and the National Institute of Occupational Safety & Health/OSHA *Pocket Guide to Chemicals*, to name a few. These guides can provide important basic information to emergency responders and are easily brought to the emergency scene.

A number of companies and government agencies maintain libraries that contain complete reference works, often consisting of full sets of material-safety data sheets, emergency plans, and environmental references. While these libraries are full of useful information, they are impractical for use during emergencies as they cannot be easily accessed by responders at the scene.

Emergency response information centers are available to emergency responders via telephone. They can respond to inquiries by searching paper/microfiche files and/or contacting predesignated industry representatives. The most popular center is CHEMTREC, which is operated by the Chemical Manufacturers Association. CHEMTREC maintains information on more than 55,000 products. In addition to providing important chemical information, it will also contact regional industry response teams to assist in the emergency response.

There are also a number of vendors of commercial databases that contain detailed information about chemicals.

Information Problems during Emergencies

Due to the availability of the aforementioned references and other information resulting from SARA and other legislation, a lot more information is now available to emergency responders. With more information becoming easier to access, the problem of responding to emergencies is rapidly changing from having too little information a few years ago, to having too much information and not enough expertise to effectively utilize it.

Having information about the chemicals involved is simply not enough. One needs to know the best spill response strategy to employ, determine what public protective actions should be implemented, where needed resources can be obtained, and who should be notified. Training can certainly help responders better integrate and analyze the information, but not even extensive, costly training can ensure that all necessary information is properly analyzed under emergency conditions. This is not to say that there are not a number of good public and private emergency response teams. Many local fire departments are training sophisticated "hazmat" teams; the EPA has technical assistance teams in each of its regions; and many companies have established in-house emergency response teams. Often these resources are not available for one reason or another. And if these teams are available, they too must use valuable time collecting and evaluating relevant chemical data before providing response recommendations.

Possible Applications of Expert Systems to Industrial Emergencies

One solution to this need for readily accessible expertise to analyze available data, and to develop response action plans, is the use of expert systems. By using an expert system, it is possible to retrieve only the technical information relevant to the particular emergency situation. It is possible to obtain a response strategy appropriate to the particular spill, based on factors such as the terrain, meteorological conditions, and others. Expert systems can be linked to geographic information systems (GIS) to provide advice on appropriate public protection actions to implement, designate the extent of an evacuation area, plot it on a map, and specify recommended police traffic-control points. The potential for emergency expert systems applications is unlimited. Unfortunately, there are very few currently available. Typically, an expert system at an industrial facility would contain in its database an inventory of MSDS's, geographical and demographic information about the area, and response plans. It would also contain what is referred to as a knowledge base, or a base of rules concerning what to do in various spill situations. A user would answer a few questions about the spill (such as the chemical involved, size, type) and meteorological conditions, and the expert system would then evaluate the situation and provide advice on the appropriate emergency response procedures.

Benefits of Expert Systems

The strengths of a first responder expert system, over standard chemical databases and other information sources are enumerated as follow:

1. Data can be interpreted in light of the nature and location of the incident. This interpretation reflects a factual analysis of the incident and the expertise of experienced first responders in analyzing and responding to the available information.

2. Uncertainty can be accommodated.

3. Data on multiple chemicals involved in an incident can be compared and evaluated to determine which are most toxic, the suit level required for cleanup, chemicals' interactions, combustion products, evacuation plans, etc.

4. Data are presented in both a raw and an interpreted form for use by responders.

5. Dispersion models are incorporated to predict downwind exposure depending on release rates and meteorology.

6. One does not have to access all of the existing chemical databases, most of which are not user-friendly.

7. The expert system can, upon request, report the logic it used to reach its conclusions/recommendations.

8. The system can include data that are normally available only from multiple sources (on-line and manual).

   In addition, expert systems are easy to use by computer novices and require little training. As noted earlier, expert systems also do not jump to conclusions nor try to defend bad decisions in the face of contrary evidence. Their decisions are not biased; they are detail-oriented and consider all possible solutions; they do not have bad days; and they do not get tired. In summary, expert systems can help improve the response to industrial emergencies by providing clear, accurate, consistent advice in a nonbiased manner. These systems never "lose their cool" under pressure and can be great tools to industrial emergency response personnel.

20

Auditing the Emergency Preparedness Program

INTRODUCTION

Emergency preparedness is a rapidly evolving, extremely important element of a corporation's overall risk management and loss control program. Since the tragic events in Bhopal, Kanawha Valley, and Chernobyl, many companies have quickly developed or enhanced their emergency preparedness programs. Emergency preparedness programs include all those activities that are necessary to prepare an organization to respond to and to mitigate the effects of an emergency or disaster. These activities include emergency planning, training, drills and exercises, community awareness, emergency resource management, and others. Without effective emergency preparedness, no company can be assured that its response will be adequate during an emergency.

Due to the recent heightened recognition of the importance of emergency preparedness, several professional trade associations, such as the Chemical Manufacturers Association, have developed guidance for their members concerning how to develop and implement an emergency preparedness program. Some of these programs include an aggressive schedule for implementation that, in certain cases, may result in the omission of critical components. And as with any new corporate program, not all plants will fully implement the program to the degree that corporate managers would like.

This chapter is intended to help corporate managers verify facility compliance with applicable emergency preparedness regulations and company policies. An emergency management audit could be made part of a corporation's environmental audit program (if one exists) or used as a stand-alone program. However it is used, it should be remembered that the audit suggested in this book differs from other types of audits (e.g., financial audits) in that there are no generally accepted principles or standards for such an emergency preparedness audit. Therefore, while this manual suggests several elements that might be included in an emergency preparedness program, it may not be all-inclusive. Also, some elements suggested here may be inappropriate for your facility.

REASONS TO AUDIT THE EMERGENCY PREPAREDNESS PROGRAM

Each company and plant should audit its emergency preparedness program for different reasons and to derive different benefits. The most basic reason for auditing an emergency preparedness program is one of common sense. Since management realizes that not all accidents or emergencies will be averted due to loss prevention programs, effective emergency response programs must be in place. The way to ensure that a capable response mechanism is in place is to have an effective emergency preparedness program in place.

BENEFITS OF AN EMERGENCY PREPAREDNESS PROGRAM AUDIT

As with other types of audits that a company might undertake, there are a number of benefits that can be derived from an emergency preparedness program audit. For example, financial audits confirm whether a company's financial management system is operating in accordance with generally accepted accounting principles and can also lead to improved corporate management. Environmental audits can help management decide if its programs comply with applicable environmental legislation and can also point out potential trouble spots in its processing systems.

Likewise, audits of an emergency preparedness program can result in numerous benefits:

Identification of potential risk. Since any sound emergency preparedness program should be based on a firm understanding of potential risks or hazards, an audit program can help to ensure that analysis of these risks or hazards occur on a regular basis. Such an analysis can lead to process changes that can help prevent certain accidents and emergencies altogether.

Increased emergency response effectiveness. As has been stated, the purpose of an emergency preparedness program is to ensure that if an emergency does occur, the response is effective. All the activities of an emergency preparedness program should contribute to this goal. If an element is deficient, then the response to an emergency may be deficient and costly. Therefore, an audit of emergency preparedness capabilities will help ensure the adequacy and effectiveness of emergency response. This of course will lead to reduced costs in terms of death, injury, environmental damage, liability, and property damage.

Management comfort or security. An emergency preparedness audit program can help management feel secure about its emergency response capabilities. This comfort stems from knowing that everything that can be done to ensure an effective response capability has been done. While this benefit is not quantifiable, it helps ensure management that its legal and ethical responsibilities are being met.

Improved community relations. An emergency preparedness audit program ensures that plant and company management is vigilant in its approach to emergency management. There can be no worse or more insecure feeling on the part of government officials and the community at large, after seeing a company pour tremendous resources into a CAER (Community Awareness and Emergency Response Program, promoted by the Chemical Manufacturers Association) type program, only to let it be forgotten shortly thereafter. Since this audit will lead to constant and continual improvement of emergency response capabilities, the public at large will be more comfortable with the company.

The reasons for developing an emergency preparedness audit program vary for each company and plant. Generally there are several legal, financial, public relations, safety, and ethical reasons for implementing such a program. The benefits derived from this program are as varied, and some benefits are clearly quantifiable while some are not. The most important benefit of assessing emergency preparedness programs stems from the fact that emergency preparedness helps ensure an effective response to emergency situations. And in the end, an effective response may be the only way to prevent an emergency from becoming a disaster.

Once it is decided that an emergency preparedness audit system would be useful, determining how best to establish one is the next step. There is no one best audit program that should be implemented, since every company is different and has different needs. The differences between these programs are based on differing goals, scope, staffing, and procedures. Some firms might run their programs out of their operating divisions, while others may do so out of their environmental or auditing departments. Some may use plant or division personnel to conduct the audits (in between their other duties), while other companies may employ full-time auditors. Some companies may preannounce their audits, while other use the element of surprise. Methods used to communicate, validate, and follow-up audit results are equally diverse. In summary, since every company's organization, operations, and management style is different, its emergency preparedness audit program must be tailor-made.

SELECTING AN APPROPRIATE AUDIT PROGRAM

The type of audit program chosen by a company will depend on the following:

- audit objectives
- available resources and personnel
- company structure and size
- potential impact on the environment
- existing environmental management efforts

Audit Objectives

A firm's audit objectives will vary with its management philosophy and size. The objectives chosen affect the scope and structure of the auditing system adopted. Some typical objectives are as follow:

1. To improve the emergency response performance of company facilities and their personnel.
2. To increase corporate-wide awareness of emergency preparedness programs.
3. To accelerate corporate-wide development of emergency preparedness and response systems.
4. To ensure compliance with applicable federal, state, and local laws.
5. To improve corporate risk management/loss prevention programs.
6. To protect companies from liability due to failure to develop sound emergency preparedness programs.

The objectives for your facility may include one or more of the preceding or may be totally different. Whatever the objectives are, they will certainly affect the design and performance of the audit. For example, an audit whose objective is to determine compliance with a certain local law will likely not be as extensive as an audit designed to improve the emergency response performance of all corporate facilities.

Available Resources and Personnel

Given unlimited resources, a company can design an audit that achieves all of the objectives previously listed. But extensive audits require a significant amount of time, money, and manpower, and are probably unwise. Since there are few (if any) companies that have unlimited resources, the structure and extensiveness of the audit program must be based on the actual availability of company resources. Companies need to look at a variety of approaches and methodologies that will allow achievement of the desired objectives, but at a reasonable cost.

Company Structure and Size

A company's structure and size affect the type of audit that will be performed. For example, decentralized companies may prefer to rely on plant personnel to carry out audits while relying on corporate personnel only in an advisory capacity. However, companies with a highly structured hierarchy are likely to utilize dedicated, full-time personnel at the corporate level for the audit. Also, while large companies tend to have the capability to allow a number of personnel to participate in the audit program, small companies can ill afford to dedicate many individuals to the effort.

Potential Impact on the Environment

Companies in different industries each impact the environment in a variety of ways. The potential impact of a company's hazards must be considered when choosing an audit system. This impact should be measured in terms of the potential effects on human life, the natural environment, and property. A small wastewater treatment facility may present certain risks to a community that a large chemical plant would not. Yet while both threats are real, the degree of required preparedness is different, and therefore the type and scope of the audits will probably be different.

Existing Environmental Management Efforts

Few companies to date have developed environmental auditing programs, let alone extensive emergency preparedness audit programs. Some companies may cover emergency preparedness, to a small degree, during their environmental audits. Many companies have only recently begun development of comprehensive emergency preparedness programs and are probably only now ready for a complete audit. Whatever the history of emergency preparedness program audits, it is important to build on the experience of previous environmental auditing and safety auditing programs. The lessons learned in these programs will be useful in the design and development of the emergency preparedness audit program. Building on past programs of a similar nature is more cost-effective than creating a system from scratch.

MANAGEMENT SUPPORT

Top management support for an emergency preparedness audit program is essential. First, this type of program requires the full participation of all plant personnel, and only genuine directives from management can ensure this on a regular basis. Various corporate initiatives are often started with lots of enthusiasm and zeal, only to be forgotten and left to decay shortly thereafter. Emergency preparedness is no different. If a disaster occurs in or near a plant, there is increased attention to emergency preparedness. If another company in your industry is struck by disaster, there is usually additional activity in this area at your facility. But as headlines fade, all too often, emergency preparedness efforts fade as well. Therefore, one way of keeping the level of activity high is through an audit program with management's firm backing.

Management can also help to ensure that the audits are conducted in an independent manner. This is the only way to ensure that the audit's objectives are achieved and the the results are meaningful. Finally, management support is essential to ensure that when the audit reveals deficiencies or noncompliance, corrective actions are implemented in timely manner.

ADOPTING AUDIT TOOLS AND PRACTICES

An emergency preparedness audit program can make use of a range of tools or procedures. Those experienced with environmental or safety audits should be familiar with the tools discussed in this section. Typically, an audit manual is prepared to guide company personnel in conducting the audit. The manual will list the applicable regulations that the facility must comply with as well as corporate policies expected to be followed. It will also state how they are to be met. Checklists will often augment the manual by the following:

- identifying records and documents to be reviewed
- asking questions
- inspecting resources and equipment
- reviewing related programs

The manual can also include a schedule that specifies the date and times for audits of specific plants, or a general time frame within which the audit will be performed. Record-keeping and reporting procedures should also be included to ensure consistency in the recording of information collected during the audit and the reporting of the results at the conclusion of the audit. Careful preparation of the audit manual will help ensure that the right information is collected, that it is methodically reviewed, and that the results are presented to management in a clear concise manner for corrective action.

Reviewing and Following up Audit Results

The discovery of a problem without subsequent correction is likely to increase the chance of corporate or individual liability. Therefore it is very important that procedures are developed to ensure that each problem identified in the audit is corrected. The types of problems likely to be discovered will vary in magnitude and importance. Regardless of the type of problem, procedures must be in place that help to identify potential solutions, choose among them, and implement them within appropriate time frames.

WHO SHOULD CONDUCT THE AUDIT?

Every organization will conduct its audit differently depending on its own specific needs, but knowing what sort of commitment is required of the audit staff is extremely important to the success of any audit program. If a firm and dedicated commitment to the audit program is not made, then it will probably not yield its intended benefits. When determining the availability of the audit staff and who shall conduct the audit, certain questions are raised: Should there be both a full-time emergency preparedness audit manager and staff, or should there be a full-time emergency preparedness audit manager with staff assigned only on a collateral basis? Or should both the emergency preparedness audit manager and staff be assigned on a collateral basis?

In deciding which option is best for your organization, you must consider a number of factors, including: (1) financial resources to support the program, (2) staff expertise, and (3) desirability for independent appraisal. A small organization is likely unable to afford a full-time audit manager and staff, but a large corporation with many plants may find it cost effective to dedicate a staff for this purpose. The use of an external consultant may be the preferred choice if adequate expertise is not available within the organization or if an independent appraisal is desired.

AUDIT TEAM CHARACTERISTICS

The quality of any type of audit is often only as good as the individuals conducting the audit. Deciding who are the best individuals to conduct an audit will vary depending on who you ask. For an audit of a particular plant's emergency preparedness program, plant managers often feel their own facility people would be the best choice to make up the audit team while outside consultants or corporate auditors would be the worst. Yet corporate mangers who feel an independent appraisal is important often feel that the best choice would be outside consultants or corporate auditors and the worst choice would be facility personnel. This points out the problems with selecting a qualified audit staff.

The answer to this problem lies in the objectives of the audit. If a plant manager desires an internal audit of the emergency preparedness program prior to, say, an OSHA inspection, then internal personnel would probably be adequate. But if the chairman of the board of directors wants the program assessed, then a more independent staff would be appropriate.

Regardless of where the staff comes from, it is important that members of the audit team have certain areas of expertise and experience, including the following:

- knowledge of applicable laws and regulations
- relevant technical expertise
- knowledge of facility operations
- knowledge of management systems
- previous assessment/audit experience
- knowledge of similar facilities' emergency programs

In addition to the factors just cited, the audit team leader should be highly experienced and be able to command the respect of other team members. The team leader must often do a difficult balancing act between ensuring that a thorough audit is done and not interfering with the plant's productivity. The team leader must be credible to management and also be able to alleviate the fears of employees who are uneasy about being audited.

TRAINING

Despite having chosen a well-qualified audit staff, audit staff training is critical. Before beginning the audit, the staff should be trained in the company's established auditing practices. Each team member should be totally familiar with the objectives of the audit, the audit plan, the approach for the upcoming audit, and his or her responsibilities. On-the-job training may also be required and desirable for new members of the team. The more experienced the staff is at assessing emergency preparedness programs, the less time will be required for training.

PREAUDIT ACTIVITIES

An emergency preparedness program audit involves a number of important steps, some of which are to be performed in advance of the on-site review. Planning and preparing for an emergency management audit can be a time-consuming, frustrating experience or can be a productive exercise that proves beneficial during the conduct of the audit. The extent of preaudit planning often depends on the experience of the assessors and on the particular facility's regulatory requirements.

The steps that should be performed in advance of the on-site audit are summarized as follow:

Step 1:   Selecting facilities to be audited.
If you are responsible for only one facility, then this step is unnecessary except for deciding on the frequency of the audit (i.e., annually, biannually, etc.). If you are responsible for more than one facility, then this is the first of your preaudit activities. There are a number of methods that can be used for determining which facilities should be assessed. For example, random selection, perceived hazards, or the importance of the facility in terms of business consideration, are all valid considerations for choosing which facility or facilities should be assessed. It is important to keep in mind that most facilities can benefit from such as audit, and all should be assessed on a regular basis even though some may require more frequent audits than may others.

Step 2:   Scheduling the audit.
Scheduling the audit may not be as simple as one might think. There are several considerations in scheduling the audit, such as scheduling the visit to the facility, selecting the audit team members, and confirming the arrangements. This step may not be all that time-consuming if the audit is to be performed by on-site personnel. But if the audit is to be done by corporate personnel or an outside consultant, then greater lead time is required. When scheduling the on-site visit, pick a week during which the plant is operating under normal conditions and when key personnel will be present. After the arrangements have been agreed to, send the facility's manager a memorandum confirming the details (i.e., scope, date, time, and place). Be sure to select audit team personnel well in advance of the audit to ensure their availability. If any member of the audit team is new, allow adequate time for training. Also arrange for replacements in case of some problem with an original team member.

Step 3:   Gathering required information.
As the leader of the audit team, you should next identify the types of information you'll require in advance of the audit. Most of this information should be available from the facility, however, some information may be available elsewhere. Facility managers are generally happy to fulfill your requests for information but are easily annoyed when the audit team does not read the information prior to the facility visit. This information should enable audit team members to ask reasonably intelligent questions. A thorough review of background information should also prevent the omission of an important question or element for audit.

Step 4:   Developing the audit plan.
This plan outlines what needs to be done during an emergency management program audit, how the audit is to be done, who will do it, and when it will be done. The plan generally includes a list of priority topics to be assessed that are chosen on the review of background information. Priority topics usually include those areas that pose a significant risk to the company, subjects found deficient in previous audits, and other elements deemed appropriate by audit team members. The plan also contains a detailed checklist (see appendix A). This checklist covers several areas necessary for an effective emergency management audit.

Step 5:   Audit team briefing.
The team leader should conduct a briefing with team members and review the purpose and scope of the audit. This can be accomplished

by reviewing the audit plan and discussing the individual assignments of team members.

POSTAUDIT ACTIVITIES

After conducting the audit in accordance with the plan, it is important that the team members provide some initial feedback to department personnel that may have been observed during the audit. This will allow department personnel to begin with implementing some obvious corrective actions but also the opportunity to correct any factual errors by team members. It is then necessary for the team leader to assemble team members for a debriefing. The comments of all team members should be reviewed and discussed by the team as a whole. Team members should then develop a report on their findings for the area they audited. These individual reports should be compiled by the team leader into a comprehensive report that identifies the strengths and weaknesses of the facility's emergency management program. It should also rank the deficiencies so that corrective actions can be taken in a prioritized manner.

The draft report should be sent to the plant manager so that he/she has an opportunity to identify factual errors in the report or to point out, in the manager's opinion, any auditor misinterpretations of fact. Any factual mistakes should be corrected prior to issuance of the final report. If there are any major disagreements between the auditors and the manager, the manager's opinion should be presented in an appendix to the report.

A typical outline for an emergency management audit report is as follows:

I. Introduction
   A. Purpose.
   B. Methodology.
II. Findings
   A. Summary of findings (by major function area, such as emergency organization, emergency plan, training, facilities and resources, and response capabilities).
   B. Recommendation for improvement (this can be broken down into a listing by priority and/or a listing by: deficiencies requiring immediate correction, areas requiring some smaller degree of corrective action, and areas that could be improved upon).

It is hoped that this report will serve as a blueprint for action by the plan in improving its emergency management program. Corrective actions should be monitored by corporate management. This report will also serve as the foundation for the next year's audit.

Conclusion

In setting out to write this book, my goal was to provide simple, practical advice on developing industrial emergency preparedness programs. While many existing guides and technical publications were used and referenced in this book, I tried to build on practical, real-world experience gained in more than nine years of planning for federal, state, local, and private organizations. Industrial emergency preparedness is a growing field and one that is essential to ensure the safety of the public and the protection of the environment. As I stated earlier in the book, the difference between emergencies and disasters is often the degree and quality of preparedness. While I hope your facility is never faced with an emergency, I believe this book can help ensure that the facility is well-prepared, just in case.

Robert B. Kelly

Emergency Management Assessment Checklist

Emergency Management Assessment Checklist

Table of Contents

Item	Provided for Yes	No	Capability Yes	No	Does Not Apply

Section 1: Basic Plan Elements

1.1 Does a comprehensive emergency preparedness plan exist? ( ) ( ) ( ) ( )

1.2 If no to 1.1 above, do written hazard specific procedures exist. (i.e., fire brigade, spill control, etc.)? ( ) ( ) ( ) ( )

1.3 Format of plan is consistent to other facility/corporate policy/programs. ( ) ( ) ( ) ( )

1.4 Statement of facility/corporate policy on emergency response included. ( ) ( ) ( ) ( )

1.5 Plan has been given authority by issuance of letter of promulgation, policy statement, cover memo, etc., signed by highest management level. ( ) ( ) ( ) ( )

1.6 Plan includes:

- Table of contents ( ) ( ) ( ) ( )
- Record of changes ( ) ( ) ( ) ( )
- Introduction/purpose ( ) ( ) ( ) ( )
- Definitions of terms and abbreviations ( ) ( ) ( ) ( )
- Distribution list ( ) ( ) ( ) ( )

1.7 Plan is available to all individuals on need-to-know basis (distribution adequate). ( ) ( ) ( ) ( )

1.8 Plan establishes emergency reaction organization. ( ) ( ) ( ) ( )

1.9 Following functions/responsibilities have been clearly defined and assigned to individuals:

- Plan administration ( ) ( ) ( ) ( )
- Direction of operation ( ) ( ) ( ) ( )
- Coordination of support ( ) ( ) ( ) ( )
- Maintenance of plan and procedures ( ) ( ) ( ) ( )
- Review hazards regularly ( ) ( ) ( ) ( )
- Training ( ) ( ) ( ) ( )
- Drills and exercises ( ) ( ) ( ) ( )
- Inventory/Maintenance of equipment ( ) ( ) ( ) ( )
- Specific Response functions ( ) ( ) ( ) ( )
- Coordination of off-site plans ( ) ( ) ( ) ( )

1.11 Plan provides for organization of each operating shift. ( ) ( ) ( ) ( )

1.12 Alternates for key positions and clear chain of command exist. ( ) ( ) ( ) ( )

Item	Provided for Yes	No	Capability Yes	No	Does Not Apply
1.13 Plan provides for/based on assessment of hazards.	( )	( )	( )	( )	
1.14 Plan provides for annual testing (drills and exercises) of capabilities.	( )	( )	( )	( )	
1.15 Provisions made to correct any discrepancies noted by reviews and operations.	( )	( )	( )	( )	
1.16 Plan is comprehensive and includes:					
- Provisions for prevention	( )	( )	( )	( )	
- Provisions for response	( )	( )	( )	( )	
- Provisions for cleanup/restoration of normal operations	( )	( )	( )	( )	
1.17 Other programs supporting/supplementing plan are referenced (e.g., Hazard Communications Program).	( )	( )	( )	( )	
1.18 Schedule of implementation included.	( )	( )	( )	( )	
1.19 In addition to assigning responsibilities, plan provides for accountability of individuals.	( )	( )	( )	( )	
1.20 Plan is to identify potential emergencies and local considerations.	( )	( )	( )	( )	
1.21 Plan is to define/establish various levels of emergencies with levels of response.	( )	( )	( )	( )	
1.22 Emergency organization consistent with operational organization.	( )	( )	( )	( )	
1.23 Basic plan elements required under OSHA 1910.38(a)(2) include:					
- Emergency evacuation procedures	( )	( )	( )	( )	
- Emergency shutdown procedures	( )	( )	( )	( )	
- Employee accountability during emergencies	( )	( )	( )	( )	
- Specification of rescue and medical duties	( )	( )	( )	( )	
- Preferred means of reporting emergencies	( )	( )	( )	( )	
- Fire prevention plan	( )	( )	( )	( )	
1.24 Policy statements, etc., provide a clear-cut priority of:					
- Safety of personnel	( )	( )	( )	( )	
- Control of hazards	( )	( )	( )	( )	
- Minimize damage to property	( )	( )	( )	( )	

Item	Provided for Yes	No	Capability Yes	No	Does Not Apply
1.25 Is the format and language of the plan clear, concise, and easily understandable?	( )	( )	( )	( )	
1.26 If the facility is classified as a RCRA storage, treatment and disposal site does the emergency plan (contingency plan) format and contents meet the requirements of EPA 264 Subpart D (264.50 - 264.55) and Subpart C (264.30 - 264.37)?	( )	( )	( )	( )	

Item	Provided for		Capability		Does Not Apply
	Yes	No	Yes	No	

Section 2: Hazard/Risk Assessment

2.1 Is present response planning
based upon risk levels?   ( )   ( )   ( )   ( )

2.2 The risk assessment is to consider:

- Past incident history   ( )   ( )   ( )   ( )
- Estimates of potential   ( )   ( )   ( )   ( )
- Estimate of probability   ( )   ( )   ( )   ( )
- Local contributing factors   ( )   ( )   ( )   ( )

2.3 All types of risk must be considered:

- Natural   ( )   ( )   ( )   ( )
- Technological   ( )   ( )   ( )   ( )
- Civil Disorders   ( )   ( )   ( )   ( )

2.4 All hazardous materials are to be listed
(refer to hazard communications) including: ( )   ( )   ( )   ( )

- Generic name
- Locations used and stored
- Sources
- Quantities
- Reactivity

[To include purchased chemical products,
raw chemicals, intermediate products and
process wastes.]

2.5 Provisions made to update hazard information.   ( )   ( )   ( )   ( )

2.6 Process diagram or map used to pinpoint
hazardous materials.   ( )   ( )   ( )   ( )

2.7 Analyses includes potential adverse impact
off-site.   ( )   ( )   ( )   ( )

2.8 Systematic analysis used to evaluate
operating systems (Fault tree analysis;
Computer modeling; etc.).   ( )   ( )   ( )   ( )

2.9 Does emergency response plan provide for
ongoing assessment?   ( )   ( )   ( )   ( )

2.10 Have written procedures that describe the
actions to be taken in situations involving
special hazards been developed (as required
by OSHA 1910.156(c)(4)]?   ( )   ( )   ( )   ( )

2.11 List of hazardous materials has been reviewed

Item	Provided for		Capability		Does Not Apply
	Yes	No	Yes	No	
against OSHA Subpart Z for specific standards that apply (relating to emergencies).	( )	( )	( )	( )	
2.12 Do accident investigation procedures include near miss and "property-damage" type accidents rather than just employee injuries?	( )	( )	( )	( )	
2.13 If the facility is classified as the following, does it comply with appropriate EPA-RCRA sections?					
- Generator of Hazwaste	( )	( )	( )	( )	
- Transporter (see EPA 263.30)	( )	( )	( )	( )	
- Treatment/storage (see EPA 264 Subparts C-D)	( )	( )	( )	( )	

Item	Provided for		Capability		Does Not
	Yes	No	Yes	No	Apply

Section 3: Prevention

3.1　Does the plan provide for or reference procedures for the prevention of emergency situations	( )	( )	( )	( )	
3.2　Provisions are made for the inspection or testing of critical equipment or components on a regular basis [including: tanks and containers, tank supports and foundations; fire-suppression equipment (per OSHA 1910.157;1910.158; 1910.159;1910.160); Detection-alarm systems (per OSHA 1910.164 and1910.165)].	( )	( )	( )	( )	

3.3　Procedures call for the review of all new processes and equipment for compliance with:

- State, federal, local and/or industry standards	( )	( )	( )	( )	
- Provisions of the plan	( )	( )	( )	( )	

3.4　Inspection-test procedures specify type and frequency (Type includes visual detailed monitoring, nondestructive testing).	( )	( )	( )	( )	
3.5　Have appropriate protective monitoring systems been installed on high-risk operations?	( )	( )	( )	( )	
3.6　Have control measures been implemented to minimize the amount of hazardous materials?	( )	( )	( )	( )	
3.7　Are good housekeeping procedures being used to control fire and health hazards [Ref. OSHA 1910.38(b)(3)]?	( )	( )	( )	( )	
3.8　Have security measures been implemented to prevent tampering with critical equipment?	( )	( )	( )	( )	
3.9　Are safety and health programs in use?	( )	( )	( )	( )	
3.10　Responsibility for prevention procedures are to be assigned to qualified individual(s).	( )	( )	( )	( )	
3.11　All maintenance or other programs that would deal with the prevention of emergencies should be referenced in the plan.	( )	( )	( )	( )	

3.12　Fire protection equipment is inspected per OSHA Subpart L provisions:

- Portable extinguishers are inspected monthly	( )	( )	( )	( )	
- Auto sprinkler system weekly	( )	( )	( )	( )	
- Yard hydrants and hoses monthly	( )	( )	( )	( )	
- Fire pump weekly	( )	( )	( )	( )	

Item	Provided for Yes	No	Capability Yes	No	Does Not Apply
- Tank or reservoir weekly	( )	( )	( )	( )	
3.13 Maintenance procedures for equipment and systems installed on heat producing equipment to prevent accidental ignition of combustible materials are included in fire prevention plan [Ref. OSHA 1910.38(b)(5)].	( )	( )	( )	( )	
3.14 A fire prevention plan, as required by OSHA 1910.38(b), is to contain:					
- A list of major workplace fire hazards and their proper handling and storage procedures, potential ignition sources, and their control procedures - Names or job titles of personnel responsible for control of fuel source hazards - Names of regular job title of those individuals responsible for maintenance of fire prevention equipment	( )	( )	( )	( )	
3.15 Do maintenance procedures on safety/protective devices or systems include procedures for insuring adequate backup protection or procedures for when equipment is down?	( )	( )	( )	( )	
3.16 Are provisions made to protect critical components, safety devices, protective systems, etc., from deliberate or inadvertent tampering or deactivation?	( )	( )	( )	( )	
3.17 Are contractor/vendor personnel coming on-site advised about or trained in hazards, safety procedures, and emergency procedures as well as personal protective equipment?	( )	( )	( )	( )	
3.18 Have dispersion modeling studies been conducted?	( )	( )	( )	( )	

Item	Provided for Yes	No	Capability Yes	No	Does Not Apply
Section 4: Direction and Control					
4.1 Does the plan establish an EOC, command post or other control point?	( )	( )	( )	( )	
4.2 A set location for the control point should be designated.	( )	( )	( )	( )	
4.3 Alternate location(s) are to be specified.	( )	( )	( )	( )	
4.4 Locations for the control center should be convenient.	( )	( )	( )	( )	
4.5 Locations must provide an adequate measure of protection from hazards (i.e., separate from building with high fire risk, etc.).	( )	( )	( )	( )	
4.6 The EOC must be adequately equipped with:					
- Communications equipment (including secondary means)	( )	( )	( )	( )	
- Warning means (alarms, etc.)	( )	( )	( )	( )	
- Protective equipment for staff (including first-aid kit)	( )	( )	( )	( )	
- Technical information on hazards, maps, etc.	( )	( )	( )	( )	
- Administrative needs (office supplies)	( )	( )	( )	( )	
- Food, water, toilet facilities, sleeping facilities, etc., adequate for expected length of stay	( )	( )	( )	( )	
4.7 Procedures are established and individual(s) assigned responsibility for maintaining equipment in state of readiness.	( )	( )	( )	( )	
4.8 Standard operating procedures have been developed that include:					
- Activation of center including notification of staff	( )	( )	( )	( )	
- Communications on-site	( )	( )	( )	( )	
- Communications off-site	( )	( )	( )	( )	
- Use of equipment/technical support	( )	( )	( )	( )	
- Press/public information	( )	( )	( )	( )	
- Determining magnitude of emergency	( )	( )	( )	( )	
4.9 A staff has been designated with individuals assigned following functions:					
- Policy	( )	( )	( )	( )	
- Analysis and coordination	( )	( )	( )	( )	
- Operations	( )	( )	( )	( )	
- Communications	( )	( )	( )	( )	

Item	Provided for Yes	No	Capability Yes	No	Does Not Apply
- Hazards/meteorological monitoring	( )	( )	( )	( )	
- Technical support	( )	( )	( )	( )	
- Clerical/administrative duties	( )	( )	( )	( )	
- Press/Public Information	( )	( )	( )	( )	
4.10 Provisions have been made for emergency power, lighting, utilities, etc.	( )	( )	( )	( )	
4.11 Up-to-date documents available include:					
- Telephone numbers lists					
- Letters of agreement with off-site agencies	( )	( )	( )	( )	
4.12 Provisions made to control access to center by unauthorized personnel.	( )	( )	( )	( )	

Item	Provided for Yes	No	Capability Yes	No	Does Not Apply

Section 5: Training

5.1 Does existing plan provide for emergency
response training?   ( )   ( )   ( )   ( )

5.2 Training programs exist outside of the
emergency plan that relate to:

- Hazardous materials   ( )   ( )   ( )   ( )
- Personal protective equipment   ( )   ( )   ( )   ( )
- Preventive maintenance   ( )   ( )   ( )   ( )
- Spill response   ( )   ( )   ( )   ( )
- First-aid   ( )   ( )   ( )   ( )
- Other   ( )   ( )   ( )   ( )

5.3 Does the facility have training capabilities
(training program director, etc.)?   ( )   ( )   ( )   ( )

5.4 The facility has:

- Videotape training   ( )   ( )   ( )   ( )
- 16mm film equipment   ( )   ( )   ( )   ( )
- Transparency projection   ( )   ( )   ( )   ( )
- Facility for on-site practical exercises
(fire control, spill control, etc.)   ( )   ( )   ( )   ( )

5.5 Emergency response training should be based
on specific hazards and response duties.   ( )   ( )   ( )   ( )

5.6 Plan must specify type and frequency of
training for each emergency response function. ( ) ( )   ( )   ( )

5.7 Training records kept to document date,
person, and type of training.   ( )   ( )   ( )   ( )

5.8 Course content based on duties outlined in plan. ( ) ( ) ( ) ( )

5.9 Provisions made to change content to reflect
changes in plan or hazards.   ( )   ( )   ( )   ( )

5.10 Periodic training review/retraining provided. ( )   ( )   ( )   ( )

5.11 Testing of student knowledge/skills provided. ( )   ( )   ( )   ( )

5.12 Training is coordinated with off-site personnel. ( ) ( ) ( ) ( )

5.13 New response employees trained within
reasonable length of time.   ( )   ( )   ( )   ( )

5.14 All persons responsible for training are

Item	Provided for Yes	Provided for No	Capability Yes	Capability No	Does Not Apply
identified.	( )	( )	( )	( )	
5.15 Training for all types of emergencies provided.	( )	( )	( )	( )	
5.16 Adequacy of training is reviewed during drills/exercises.	( )	( )	( )	( )	
5.17 Training methods include:					
- Formal classroom	( )	( )	( )	( )	
- Lecture instruction	( )	( )	( )	( )	
- Over-the-shoulder, hands-on instruction	( )	( )	( )	( )	
- Off-site course	( )	( )	( )	( )	
5.18 Where possible, emergency training should be incorporated with normal operating training.	( )	( )	( )	( )	
5.19 Minimum training levels for all emergency response activities are defined.	( )	( )	( )	( )	
5.20 All employees are to be trained/instructed in the following at least yearly:					
- Hazardous materials, including fire hazards [per OSHA 1910.38(b)(4) and Hazard Communications Act]	( )	( )	( )	( )	
- Evacuation procedures	( )	( )	( )	( )	
- Emergency reporting procedures	( )	( )	( )	( )	
- Fire extinguisher use [per OSHA 1910.158(g)(l)]	( )	( )	( )	( )	
- Leak/emergency warning signs (odor, smoke, sounds, etc.)	( )	( )	( )	( )	
5.21 Training of firefighting personnel is to include the following, commensurate with duties of each member [Ref. OSHA 1910.156(c) and Appendix A to Subpart L]:	( )	( )	( )	( )	
- Members assigned to interior firefighting are to be provided with a training session at least quarterly	( )	( )	( )	( )	
- All members to be trained at least annually	( )	( )	( )	( )	
- To include review of written procedures describing actions to be taken in situations involving special hazards of specific materials	( )	( )	( )	( )	
- Fire brigade officers and instructors are to be given a higher level of training	( )	( )	( )	( )	
- Quality and content of training must be similar to accepted training courses	( )	( )	( )	( )	
5.22 Training of response personnel should follow these guidelines:					

Item	Provided for		Capability		Does Not Apply
	Yes	No	Yes	No	
- Training directly based on expected situations	( )	( )	( )	( )	
- Session should be relatively short but repeated frequently	( )	( )	( )	( )	
- Skills must be practiced frequently	( )	( )	( )	( )	
- Training must be realistic	( )	( )	( )	( )	
- Everyone should participate	( )	( )	( )	( )	

Item	Provided for		Capability		Does Not Apply
	Yes	No	Yes	No	

Section 6: Drills and Exercises

6.1 Does the present plan contain provisions for the use of drills and exercises? ( ) ( ) ( ) ( )

6.2 Are drills and exercises used as a means of testing the plan or solely for training? ( ) ( ) ( ) ( )

6.3 Have there been any drills and exercises within the last year? ( ) ( ) ( ) ( )

6.4 Is there an annual drill or exercise that tests the entire plans capabilities? ( ) ( ) ( ) ( )

6.5 The results of drills must be evaluated and documented. ( ) ( ) ( ) ( )

6.6 Drills are to be based on valid scenarios. ( ) ( ) ( ) ( )

6.7 Are provisions made for correcting defects in the plan that are detected by the drills/exercises? ( ) ( ) ( ) ( )

6.8 Off-site personnel/agencies participate in drills? ( ) ( ) ( ) ( )

6.9 In addition to a yearly "full-plan" drill, periodic exercises on key elements should be held on:

- Communications ( ) ( ) ( ) ( )
- Fire control response ( ) ( ) ( ) ( )
- Medical/first-aid ( ) ( ) ( ) ( )
- Spill Control ( ) ( ) ( ) ( )
- EOC/control staff ( ) ( ) ( ) ( )
- Monitoring ( ) ( ) ( ) ( )
- Cleanup ( ) ( ) ( ) ( )
- Evacuations of individual areas/full facility ( ) ( ) ( ) ( )

6.10 Various different types of exercises are used:

- Tabletop ( ) ( ) ( ) ( )
- Functional ( ) ( ) ( ) ( )
- Full-scale ( ) ( ) ( ) ( )

6.11 Scenario development includes the following elements:

- Plan review/needs analysis ( ) ( ) ( ) ( )
- Defines scope ( ) ( ) ( ) ( )
- Costs/liabilities considered ( ) ( ) ( ) ( )
- Statement of purpose ( ) ( ) ( ) ( )
- Identifies resources needed ( ) ( ) ( ) ( )
- Contains major event sequence ( ) ( ) ( ) ( )
- Defines expected response ( ) ( ) ( ) ( )

Item	Provided for Yes	No	Capability Yes	No	Does Not Apply
6.12 Does participation in drills/ exercises include all levels including top management?	( )	( )	( )	( )	
6.13 Has responsibility for developing, scheduling, and conducting drills/ exercises been specifically assigned to a responsible party?	( )	( )	( )	( )	
6.14 Has input of all levels of emergency organization been considered in exercise planning and review?	( )	( )	( )	( )	

Item	Provided for Yes	No	Capability Yes	No	Does Not Apply

Section 7: Supplies and Equipment

	Item	Provided for Yes	No	Capability Yes	No	Does Not Apply
7.1	Are provisions for determining supplies and equipment needs/availability included in the plan?	( )	( )	( )	( )	
7.2	Have provisions for an inventory list of equipment been included?	( )	( )	( )	( )	
7.3	Are inventories kept current on a regular schedule?	( )	( )	( )	( )	
7.4	Maintenance and decontamination procedures are described.	( )	( )	( )	( )	
7.5	Maintenance is based on standards of the manufacturer and others.	( )	( )	( )	( )	
7.6	Equipment is tested as specified.	( )	( )	( )	( )	
7.7	Maintenance and testing is documented.	( )	( )	( )	( )	
7.8	Maintenance and testing personnel are properly trained.	( )	( )	( )	( )	
7.9	Emergency personnel have easy access to equipment.	( )	( )	( )	( )	
7.10	Easy access by unauthorized personnel is minimized.	( )	( )	( )	( )	
7.11	Procedures for determining minimum supply levels of expendable supplies are provided.	( )	( )	( )	( )	
7.12	Supplies of personal protective equipment and other supplies for emergency response activities are kept separate from normal operating supplies.	( )	( )	( )	( )	
7.13	List of special equipment, based on hazard information on specific materials, is updated as changes in materials occur.	( )	( )	( )	( )	
7.14	Lists of emergency sources of supplies and equipment is maintained and available to response personnel.	( )	( )	( )	( )	
7.15	Available equipment includes:					
	- First-aid supplies	( )	( )	( )	( )	
	- Personal protective equipment	( )	( )	( )	( )	

Item	Provided for		Capability		Does Not Apply
	Yes	No	Yes	No	
- Communications equipment	( )	( )	( )	( )	
- Firefighting equipment	( )	( )	( )	( )	
- Spill control equipment	( )	( )	( )	( )	
- Spill cleanup equipment	( )	( )	( )	( )	
- Exposure/release monitoring devices	( )	( )	( )	( )	
- Tools for repair of hazardous equipment	( )	( )	( )	( )	
- Neutralizing agents	( )	( )	( )	( )	

7.16   Is access to equipment equal
for all shifts?   ( )   ( )   ( )   ( )

7.17   Protective equipment is provided
to emergency brigade in accordance
of OSHA regulations [1910.156(e)(l)(ii)]:
- Foot and leg protection   ( )   ( )   ( )   ( )
- Body protection   ( )   ( )   ( )   ( )
- Hand protection   ( )   ( )   ( )   ( )
- Head, eye, and face protection   ( )   ( )   ( )   ( )

7.18   Respiratory protection for firefighters
is in compliance with OSHA regulation
1910.156(f).   ( )   ( )   ( )   ( )

7.19   Respiratory equipment quantitative fit test
procedures are available for inspection as
required by OSHA 1910.156(f)(2)(111).   ( )   ( )   ( )   ( )

7.20   Is there meteorological
capacity on-site or have provisions
been made to obtain this information
from an off-site source?   ( )   ( )   ( )   ( )

7.21   If this facility is classified as a RCRA HazWaste
treatment, storage and disposal site does it
comply with section EPA 264.32 and have:
- Internal communications or alarm system   ( )   ( )   ( )   ( )
- Telephone or other two-way communications
device to off-site response agencies   ( )   ( )   ( )   ( )
- Fire control equipment   ( )   ( )   ( )   ( )
- Adequate water, firefighting foam supplies ( )   ( )   ( )   ( )

7.22   Was consideration given to problems of
communication during the selection of SCBA
equipment?   ( )   ( )   ( )   ( )

7.23   Does SCBA equipment have locator system?   ( )   ( )   ( )   ( )

Item	Provided for Yes	No	Capability Yes	No	Does Not Apply

Section 8: Off-site Resources

8.1 Have the capabilities and resources of the
following community organizations been reviewed
and considered in the appropriate response
procedures:

- Police	( )	( )	( )	( )	
- Fire	( )	( )	( )	( )	
- Emergency Management Agency	( )	( )	( )	( )	
- Public Health Dept.	( )	( )	( )	( )	
- Environmental Protection Agency	( )	( )	( )	( )	
- Dept. of Transportation Public Works	( )	( )	( )	( )	
- Water Supply	( )	( )	( )	( )	
- Sanitation	( )	( )	( )	( )	
- Port Authority	( )	( )	( )	( )	
- Transit Authority	( )	( )	( )	( )	
- Rescue Squad	( )	( )	( )	( )	
- Ambulance	( )	( )	( )	( )	
- Hospitals	( )	( )	( )	( )	
- Utilities	( )	( )	( )	( )	
- Community Officials	( )	( )	( )	( )	
- Red Cross	( )	( )	( )	( )	
- Salvation Army	( )	( )	( )	( )	
- Church Groups	( )	( )	( )	( )	
- Ham Radio Operators	( )	( )	( )	( )	
- Off-Road Vehicle Clubs	( )	( )	( )	( )	

8.2 Have the capabilities and resources of the
following State agencies been reviewed
and considered in the appropriate response
procedures:

- State Emergency Response Commission (SERC)	( )	( )	( )	( )	
- State Environmental Protection Agency	( )	( )	( )	( )	
- Emergency Management Agency	( )	( )	( )	( )	
- Department of Transportation	( )	( )	( )	( )	
- Police	( )	( )	( )	( )	
- Public Health Department	( )	( )	( )	( )	
- Department of Agriculture	( )	( )	( )	( )	

8.3 Have the capabilities and resources of the
following federal agencies been reviewed
and considered in the appropriate response
procedures:

- National Response Center	( )	( )	( )	( )	
- Agency for Toxic Substances and Disease Registry	( )	( )	( )	( )	
- Federal Emergency Management Agency	( )	( )	( )	( )	
- Federal On-Scene Coordinator	( )	( )	( )	( )	
- U.S. Department of Transportation	( )	( )	( )	( )	

Item	Provided for		Capability		Does Not Apply
	Yes	No	Yes	No	
- U.S. Coast Guard	( )	( )	( )	( )	
- U.S. Environmental Protection Agency	( )	( )	( )	( )	

8.4   Have the capabilities and resources of the
following industry organizations been reviewed
and considered in the appropriate response procedures:

Item					
- Transporters	( )	( )	( )	( )	
- Chemical Producers/Consumers	( )	( )	( )	( )	
- Spill Cooperatives	( )	( )	( )	( )	
- Spill Response Teams	( )	( )	( )	( )	
- CHEMTREC	( )	( )	( )	( )	
- CHEMNET	( )	( )	( )	( )	
- CHLOREP	( )	( )	( )	( )	
- NACA Pesticide Safety	( )	( )	( )	( )	
- Association of American Railroads	( )	( )	( )	( )	
- Poison Control Center	( )	( )	( )	( )	

Item	Provided for Yes	No	Capability Yes	No	Does Not Apply

Section 9: Mutual Aid

Item	Provided for Yes	No	Capability Yes	No	Does Not Apply
9.1 Do any mutual aid agreements, written or verbal, exist?	( )	( )	( )	( )	
9.2 Does the agreement clearly define level of/type of support to be given?	( )	( )	( )	( )	
9.3 Is a listing of capabilities for all facilities included (list to include equipment, personnel, supplies, expertise available)?	( )	( )	( )	( )	
9.4 Are conditions that would warrant assistance defined?	( )	( )	( )	( )	
9.5 Are agreements legally binding?	( )	( )	( )	( )	
9.6 Are methods of communications defined?	( )	( )	( )	( )	
9.7 Do procedures exist for coordinating response activities?	( )	( )	( )	( )	
9.8 Do clearly defined command structures exist?	( )	( )	( )	( )	
9.9 Are methods/rates of compensation defined?	( )	( )	( )	( )	
9.10 Are liabilities clearly defined?	( )	( )	( )	( )	
9.11 Have drills, exercises, and training sessions involving all parties been held?	( )	( )	( )	( )	
9.12 Have liaison personnel for each member been designated?	( )	( )	( )	( )	
9.13 Are regular planning sessions held?	( )	( )	( )	( )	
9.14 Have considerations for the protection of trade secrets and proprietary information been agreed upon?	( )	( )	( )	( )	

Item	Provided for		Capability		Does Not Apply
	Yes	No	Yes	No	

Section 10: Detection

10.1 Does the plan provide for the use of detection systems

	( )	( )	( )	( )	

10.2 Automatic detection devices are in use

	Provided Yes	Provided No	Capability Yes	Capability No	
- Smoke detectors	( )	( )	( )	( )	
- Heat detectors	( )	( )	( )	( )	
- Remote single-substance monitors	( )	( )	( )	( )	
- Leak detectors					
- Process control alarms	( )	( )	( )	( )	

10.3 Are detection devices monitored continuously?

	( )	( )	( )	( )	

10.4 Have provisions been made for the regular testing, inspection, maintenance, and calibration of devices?

	( )	( )	( )	( )	

10.5 Have procedures been developed for interpretation, response, and reporting of data from detectors or process controls?

	( )	( )	( )	( )	

10.6 Is training conducted on interpretation and response of data and on detection hazards?

	( )	( )	( )	( )	

10.7 Are regularly scheduled checks of hazard areas conducted with portable detection devices?

	( )	( )	( )	( )	

Item	Provided for		Capability		Does Not Apply
	Yes	No	Yes	No	

Section 11: Alerting and Warning

Item	Provided Yes	Provided No	Capability Yes	Capability No	Does Not Apply
11.1 Are provisions for alerting response personnel and warning of impending danger from an emergency situation included in the plan?	( )	( )	( )	( )	
11.2 Has a central point (such as the switchboard, security guard, etc.) been designated to receive reports of an emergency?	( )	( )	( )	( )	
11.3 Procedures have been developed for alerting appropriate response personnel as specified by OSHA 1910.165(b):					
- Via telephone utilizing lists	( )	( )	( )	( )	
- Via radio page(beepers)	( )	( )	( )	( )	
- Via plant intercom/page	( )	( )	( )	( )	
11.4 Has an alternate method (from above) been designated as required by OSHA section 1910.165(d)(3)?	( )	( )	( )	( )	
11.5 Has an individual been assigned responsibility for issuing a general warning via the following:					
- Plant-wide alarm	( )	( )	( )	( )	
- Coded alarm system	( )	( )	( )	( )	
- Plant intercom/page	( )	( )	( )	( )	
- Telephone	( )	( )	( )	( )	
11.6 Has a secondary method been designated as required by OSHA 1910.165(d)(3)?	( )	( )	( )	( )	
11.7 Has an alternate method been designated for remote areas without an alarm or page, etc.?	( )	( )	( )	( )	
11.8 Does the plan provide for hazard specific warnings (different alarm for fire, spill, etc.)?	( )	( )	( )	( )	
11.9 Are systems maintained according to OSHA section 1910.165(d)?	( )	( )	( )	( )	
11.10 Are periodic/regular tests conducted at least every two months according to OSHA 1910.165(d)(2)?	( )	( )	( )	( )	
11.11 The notification procedures should be:					
- Brief	( )	( )	( )	( )	
- Accessible to all who need to know	( )	( )	( )	( )	

Item	Provided for Yes	No	Capability Yes	No	Does Not Apply
- Simple	( )	( )	( )	( )	
11.12 Provisions have been made for:					
Alerting off-shift personnel	( )	( )	( )	( )	
Having on-call personnel report location	( )	( )	( )	( )	
Notification of off-site personnel/organizations	( )	( )	( )	( )	
11.13 Are alarms/signaling devices capable of being heard over ambient noise levels per OSHA 1910.156(b)(2)?	( )	( )	( )	( )	
11.14 If the same signalling method is used for fire brigade and evacuation alerting, does a means exist providing for distinguishing between each?	( )	( )	( )	( )	
11.15 If telephones are means of reporting emergency, are reporting procedures/telephone numbers posted?	( )	( )	( )	( )	
11.16 When using telephones, are emergency message given priority?	( )	( )	( )	( )	
11.17 Are provisions made to insure that all supervisory type alarms are properly supervised per OSHA 1910.165(d)(4)?	( )	( )	( )	( )	
11.18 Are provisions made to insure that only trained and designated employees perform maintenance on alarm systems as required by OSHA 1910.165(d)(5)?	( )	( )	( )	( )	
11.19 Are provisions made for notifying visitors of an emergency?	( )	( )	( )	( )	
11.20 Notification procedures should attempt to answer the following seven questions:					
- What happened?	( )	( )	( )	( )	
- Where did it happen?	( )	( )	( )	( )	
- Who did it happen to?	( )	( )	( )	( )	
- When did it happen?	( )	( )	( )	( )	
- How did it happen?	( )	( )	( )	( )	
- To what extent?	( )	( )	( )	( )	
- What help is needed?	( )	( )	( )	( )	

Item	Provided for Yes	No	Capability Yes	No	Does Not Apply
Section 12: Communications					
12.1 Are communications procedures provided for in the plan?	( )	( )	( )	( )	
12.2 Communications systems available include:					
- Two-way radios	( )	( )	( )	( )	
- Intercom/paging	( )	( )	( )	( )	
- Telephone	( )	( )	( )	( )	
- Runners (verbal or written messages)	( )	( )	( )	( )	
12.3 Provisions for communications exist between:					
- EOC/control and response teams	( )	( )	( )	( )	
- Response team to response team	( )	( )	( )	( )	
- EOC/control with all off-site agencies	( )	( )	( )	( )	
- EOC/control and support personnel (including press/public relations)	( )	( )	( )	( )	
- Technical support (e.g., CHEMTREC)	( )	( )	( )	( )	
12.4 Backup systems determined.	( )	( )	( )	( )	
12.5 Equipment is maintained and tested regularly.	( )	( )	( )	( )	
12.6 Procedures are tested regularly.	( )	( )	( )	( )	
12.7 Adequate communications has been provided for at the alternate EOC/ control site.	( )	( )	( )	( )	
12.8 Operating procedures have been developed including:					
- Assignment of all operational functions	( )	( )	( )	( )	
- Development radio protocol including codes ("10" codes, etc.)	( )	( )	( )	( )	
- System of prioritizing messages	( )	( )	( )	( )	
- System of tracking	( )	( )	( )	( )	
- System of documenting messages	( )	( )	( )	( )	
12.9 Has consideration been made as to the means of communicating between response team members when using SCBA respirators?	( )	( )	( )	( )	

Item	Provided for Yes	No	Capability Yes	No	Does Not Apply
Section 13: Emergency Response Teams/Organizations					
13.1 The plan provides for emergency response forces to deal with:					
- Fires	( )	( )	( )	( )	
- Explosions	( )	( )	( )	( )	
- Chemical spills	( )	( )	( )	( )	
- Atmospheric releases	( )	( )	( )	( )	
- Other	( )	( )	( )	( )	
13.2 Is each team specially trained and equipped to respond to the specific emergency?	( )	( )	( )	( )	
13.3 Have specific tasks been assigned to individuals?	( )	( )	( )	( )	
13.4 Are minimum manpower requirements (based on tasks to be performed) established?	( )	( )	( )	( )	
13.5 Are up-to-date rosters available?	( )	( )	( )	( )	
13.6 Has a responsible individual been appointed as a supervisor (for each team and on each shift)?	( )	( )	( )	( )	
13.7 Is attendance at training, drills and meetings mandatory?	( )	( )	( )	( )	
13.8 Is membership voluntary?	( )	( )	( )	( )	
13.9 Is compensation given to team members?	( )	( )	( )	( )	
13.10 Standard operating procedures have been established that include:					
- Clear definition of chain of command	( )	( )	( )	( )	
- Method of responding	( )	( )	( )	( )	
- Breakdown of tasks	( )	( )	( )	( )	
- Coordination points with other facility personnel	( )	( )	( )	( )	
- Coordination points with off-site organization	( )	( )	( )	( )	
13.11 Training, drills, etc., includes participation of off-site organization.	( )	( )	( )	( )	
13.12 A priorities list exists as to what emergency response activities are first, with first-aid and rescue at the top.	( )	( )	( )	( )	
13.13 Techniques and skills that are taught are responsive to the special hazards and available equipment.	( )	( )	( )	( )	

Item	Provided for		Capability		Does Not Apply
	Yes	No	Yes	No	
13.14 Spill control procedures include:					
- Determining what debris is to be removed	( )	( )	( )	( )	
- Preplanned disposal sites and transportation	( )	( )	( )	( )	
- Designate proper containers for storage and transport	( )	( )	( )	( )	
13.15 Response personnel are required to wear all personal protective equipment required by OSHA 1910.156e and (f).	( )	( )	( )	( )	
13.16 All personnel assigned as response personnel are physically able to perform such work [as specified by OSHA 1910.156(b)(2)].	( )	( )	( )	( )	
13.17 Have procedures been established to provide for accountability of response personnel when entering/ leaving a "hot" zone?	( )	( )	( )	( )	
13.18 Do response procedures for measuring and marking "hot" zone and controlling access exist?	( )	( )	( )	( )	
13.19 Have procedures been implemented for determining chemical substances involved in any situation immediately?	( )	( )	( )	( )	
13.20 Have guidelines been established for the selection/use of the proper protective equipment required by the particular substance(s) or hazards?	( )	( )	( )	( )	
13.21 Do rescue plan procedures include:					
- Procedures for the use of backup personnel specially equipped for entry into a "hot" zone	( )	( )	( )	( )	
- Preplanned provisions for routes of entry and at least two escape routes	( )	( )	( )	( )	
- Means of marking routes(e.g., use of rope or tape) to assist entry/exit from interior when visibility is impaired	( )	( )	( )	( )	
- Providing ladder, etc., to aid access to otherwise free areas by personnel wearing protective equipment that will hinder mobility	( )	( )	( )	( )	

Item	Provided for Yes	No	Capability Yes	No	Does Not Apply

Section 14: Facility Evacuation

Item	Provided for Yes	No	Capability Yes	No	Does Not Apply
14.1 The plan should include provisions for the emergency evacuation of personnel.	( )	( )	( )	( )	
14.2 Evacuation procedures for each areas as well as the entire facility are to be established.	( )	( )	( )	( )	
14.3 At least two (a primary and secondary) evacuation routes should be planned from each area.	( )	( )	( )	( )	
14.4 A prearranged alarm or signal is to be used to inform employees.	( )	( )	( )	( )	
14.5 A responsible individual is to be assigned authority to declare a full evacuation.	( )	( )	( )	( )	
14.6 The order to return ("All clear") is to be given only by an authorized individual.	( )	( )	( )	( )	
14.7 Individuals in each area are to be assigned certain tasks:					
- Guide others to evacuation route	( )	( )	( )	( )	
- Check area for strangers	( )	( )	( )	( )	
- Turn off noncritical equipment, close windows, doors, etc.	( )	( )	( )	( )	
14.8 Does the plan provide for at lease one drill per year?	( )	( )	( )	( )	
14.9 Has there been an evacuation drill within the past 12 months?	( )	( )	( )	( )	
14.10 Are exits in facility marked?	( )	( )	( )	( )	
14.11 Are all employees instructed in evacuation procedures?	( )	( )	( )	( )	
14.12 Are maps and/or instructions posted?	( )	( )	( )	( )	
14.13 Have assembly areas been designated with regard to safe distances, etc.?	( )	( )	( )	( )	
14.14 Provisions have been made to account for all individuals:					
- Instructions established as to where to report if an individual is not in normal area	( )	( )	( )	( )	
- Procedures for each area to account for and report on presence of area personnel	( )	( )	( )	( )	

Item	Provided for Yes	No	Capability Yes	No	Does Not Apply
- Means provided to account for visitors	( )	( )	( )	( )	
- Member of control center responsible for accounting	( )	( )	( )	( )	
14.15 Does the plan provide for special procedures for handicapped?	( )	( )	( )	( )	
14.16 Have provisions been made for temporary shelter or transport?	( )	( )	( )	( )	

Item	Provided for Yes	No	Capability Yes	No	Does Not Apply

Section 15: Security Considerations

15.1 Does the plan consider/define the role of the security function?

( )   ( )   ( )   ( )

15.2 The plan has a security force:

- In-house   ( )   ( )   ( )   ( )
- Contract   ( )   ( )   ( )   ( )

15.3 Are provisions made to control access to the facility during an emergency?

( )   ( )   ( )   ( )

15.4 Procedures have been established to control to key areas:

- EOC/control point   ( )   ( )   ( )   ( )
- Media Center   ( )   ( )   ( )   ( )
- Emergency supplies storage ( )   ( )   ( )   ( )

15.5 Have adequate precautions been taken to prevent tampering with emergency equipment (e.g., sprinkler, values)

( )   ( )   ( )   ( )

15.6 Have provisions been made for the control of traffic in and around the facility?

( )   ( )   ( )   ( )

15.7 Are procedures in place for the control of pilferage during and after an emergency?

( )   ( )   ( )   ( )

15.8 Have areas/items of high security risk (i.e., proprietary materials) been identified?

( )   ( )   ( )   ( )

15.9 Are security personnel utilized for preventative functions (e.g., inspection of fire extinguishers)?

( )   ( )   ( )   ( )

15.10 Physical security devices exist:

- Fences   ( )   ( )   ( )   ( )
- Alarms   ( )   ( )   ( )   ( )
- Locking barriers ( )   ( )   ( )   ( )

15.11 Have provisions been made for access of secured areas by emergency personnel during an emergency?

( )   ( )   ( )   ( )

Item	Provided for		Capability		Does Not Apply
	Yes	No	Yes	No	

Section 16: Public Relations

16.1 Does the plan include public relations functions before, during, and after an emergency?	( )	( )	( )	( )	
16.2 Is there an existing facility's public relations function?	( )	( )	( )	( )	
16.3 The plan calls for the development of a public information document(s) containing the following:					
- Hazard information	( )	( )	( )	( )	
- Process/facility description	( )	( )	( )	( )	
- Products produced (keyed to consumers use)	( )	( )	( )	( )	
- Safety record	( )	( )	( )	( )	
- Emergency plan summary	( )	( )	( )	( )	
- Environmental protection policy	( )	( )	( )	( )	
- Contributions to community	( )	( )	( )	( )	
- Annual report/financial information	( )	( )	( )	( )	
16.4 Information distribution method used:					
- Through employees	( )	( )	( )	( )	
- News media	( )	( )	( )	( )	
- General public through civic/community groups	( )	( )	( )	( )	
- General distribution to public	( )	( )	( )	( )	
16.5 Is upper management utilized in information program?	( )	( )	( )	( )	
16.6 Does program utilize off-site assistance (government and civic)?	( )	( )	( )	( )	
16.7 Is the same personnel responsible for information program and public information during an emergency?	( )	( )	( )	( )	
16.8 Are P.R. personnel fully trained in response plan?	( )	( )	( )	( )	
16.9 Are contacts with media established and maintained?	( )	( )	( )	( )	
16.10 Plan specifies yearly review and update of information.	( )	( )	( )	( )	
16.11 Has an operations plan been established for the control of public information during an emergency?	( )	( )	( )	( )	

Item	Provided for Yes	No	Capability Yes	No	Does Not Apply
16.12 Do the procedures specify the establishment of a press area?	( )	( )	( )	( )	
16.13 Public relations should be included in first call out group.	( )	( )	( )	( )	
16.14 Provisions requiring EOC/ control function to keep P.R. personnel up to date are included.	( )	( )	( )	( )	
16.15 General policy on information to be released should be established.	( )	( )	( )	( )	
16.16 All press/public inquires should be logged.	( )	( )	( )	( )	
16.17 Names and information of dead and injured are restricted.	( )	( )	( )	( )	
16.18 During emergency situation regularly scheduled releases should be made.	( )	( )	( )	( )	
16.19 Is bilingual capacity necessary?	( )	( )	( )	( )	

Item	Provided for Yes	No	Capability Yes	No	Does Not Apply

Section 17: Coordination Between Facility and Off-site Organizations

17.1 Does the plan provide for coordination between the facility and off-site organizations?   ( )   ( )   ( )   ( )

17.2 Do written agreements exist?   ( )   ( )   ( )   ( )

17.3 Develop a listing of outside agencies/organizations with emergency response capabilities.   ( )   ( )   ( )   ( )

17.4 Contacts with organizations having needed capabilities have been established.   ( )   ( )   ( )   ( )

17.5 Support groups (such as fire and police) have been familiarized with facility.   ( )   ( )   ( )   ( )

17.6 A list of contacts has been made showing information to be given at time of emergency.   ( )   ( )   ( )   ( )

17.8 List of off-site support includes contractors and vendors.   ( )   ( )   ( )   ( )

17.9 Regularly scheduled meetings are held to discuss support activities.   ( )   ( )   ( )   ( )

17.10 Written operating procedures exist that include:

- Communications   ( )   ( )   ( )   ( )
- Type of support   ( )   ( )   ( )   ( )
- Compensation   ( )   ( )   ( )   ( )
- Liability limits   ( )   ( )   ( )   ( )
- Contacts   ( )   ( )   ( )   ( )
- Circumstances that would require assistance   ( )   ( )   ( )   ( )
- Technical/hazard information   ( )   ( )   ( )   ( )

17.11 Off-site personnel included in training.   ( )   ( )   ( )   ( )

17.12 Procedures tested by drills and exercises.   ( )   ( )   ( )   ( )

17.13 Contact with off-site organizations includes consulting physicians, hospital, and public health organizations.   ( )   ( )   ( )   ( )

Item	Provided for		Capability		Does Not Apply
	Yes	No	Yes	No	
Section 18: Emergency Shutdown Procedures					
18.1 Does the emergency plan provide for the emergency shutdown of operating systems?	( )	( )	( )	( )	
18.2 Have specific individuals been designated to be responsible for shutdown implementation?	( )	( )	( )	( )	
18.3 Have procedures checklists been developed for each individual operations, equipment, or area?	( )	( )	( )	( )	
18.4 Are checklists immediately available to both operating and response personnel?	( )	( )	( )	( )	
18.5 Are provisions made for the availability of special tools (e.g., fire hydrant, wrench) for shutdown including backups?	( )	( )	( )	( )	
18.6 Have those operations, valves, etc., that must not be shutdown been determined?	( )	( )	( )	( )	
18.7 Have operations that require time for full shutdown been determined and time required shown	( )	( )	( )	( )	
18.9 Have diagrams, maps been made showing critical components	( )	( )	( )	( )	
18.10 Are critical components, valves, and controls clearly identifiable by color coding or marking	( )	( )	( )	( )	
18.11 Have provisions been made to identify individuals with special technological knowledge on each operational component and make them available to emergency personnel?	( )	( )	( )	( )	
18.12 Shutdown procedures should include details as to shutting of doors and windows, securing of documents and equipment, turning off lights, and locking doors as appropriate.	( )	( )	( )	( )	

Item	Provided for		Capability		Does Not Apply
	Yes	No	Yes	No	

Section 19: Recovery Planning

19.1 Does the plan include provisions for disaster recovery? ( ) ( ) ( ) ( )

19.2 Are there provisions for a clear line of succession for all top management and key personnel? ( ) ( ) ( ) ( )

19.3 Does the plan designate an alternate location for functioning of operations management? ( ) ( ) ( ) ( )

19.4 Has a recovery control team been established? ( ) ( ) ( ) ( )

19.5 Has a resource list been developed showing sources of replacement equipment (to purchase or rent), contractors, etc.? ( ) ( ) ( ) ( )

19.6 Have agreements been made with other facilities for time sharing of equipment or facilities? ( ) ( ) ( ) ( )

19.7 Have procedures, consistent with insurance careers requirements, for accounting and documenting all compensible losses been established? ( ) ( ) ( ) ( )

19.8 Have procedures been developed for preserving accident scene for accident investigations by on-site or off-site personnel? ( ) ( ) ( ) ( )

19.9 Maintain a list of contacts with insurance carriers, corporate, etc., that are to be contacted. ( ) ( ) ( ) ( )

19.10 Procedures should be developed to include:

- Assigning personnel to supervise cleanup and repair ( ) ( ) ( ) ( )
- Maintaining list of personnel for call in ( ) ( ) ( ) ( )
- Notification procedures to inform personnel not to report to work as scheduled ( ) ( ) ( ) ( )
- Damage-assessment list of necessary repairs/ replacement to be prioritized ( ) ( ) ( ) ( )
- Special procedures to expedite issuance of work orders, purchase orders, etc. ( ) ( ) ( ) ( )
- Maintaining list quantities and locations of available clean up equipment ( ) ( ) ( ) ( )
- Designating area for temporary storage of damaged equipment and materials until released by insurance investigation ( ) ( ) ( ) ( )
- Special accounting procedures to insure accurate loss figures ( ) ( ) ( ) ( )
- Provisions for accident investigation and

Item	Provided for		Capability		Does Not Apply
	Yes	No	Yes	No	
response critique to be used to upgrade plan	( )	( )	( )	( )	
- Procedures for measuring level of contamination and safe levels for reentry (including designating responsible individuals).	( )	( )	( )	( )	

Appendix B

Sample Emergency Management Plan

Table of Contents

1.0   BASIC PLAN

1.1   SUBJECT: Introduction

The COMPANYNAME is located in an area vulnerable to major hazards which may result from natural or human caused phenomena. For this reason, the COMPANYNAME has developed this plan in order to be prepared to respond to these emergencies.

In recognition of the fact that many elements of the emergency response to all hazards are similar, and whereas there are almost limitless contingencies which could be encountered, this plan has been designed to be applicable to any emergency that may affect the COMPANYNAME. However, the plan contains certain hazard-specific information in cases where the hazard requires a response that is unique and not covered by other elements of the plan.

Also, this plan is more than just a response plan. It is based on the concept that one must do more than simply respond to emergencies, that the only way to effectively manage emergencies is to plan to prevent, prepare for, respond to, and recover from emergencies. This concept is known as Comprehensive Emergency Management and is fully implemented in this plan.

All members of the COMPANYNAME staff and its management are to abide by the provisions of this plan and are to participate in the plan's development and administration. Only through a concerted and timely effort can the full benefits and goals of this plan be attained.

1.2   SUBJECT: Purpose, Policy, and Legal Authority

Purpose:

This document has been developed to provide an organizational and procedural framework for the management of emergency incidents that affect the plant. The plan also provides for the coordination between the COMPANYNAME and government for the further protection of COMPANYNAME employees and property, as well as that of the surrounding community and environment.

Policy:

It is the policy of the COMPANYNAME that its employees, property, environment, and the general public be protected from any harm that may occur as a result of its operations. This plan is being implemented to comply with the overall COMPANYNAME policy of providing its workers with a workplace free from recognized hazards.

This plan is designed to comply with the policy that all potential hazardous conditions remain within plant boundaries and will not represent a threat to the health and safety of the general public. It is also the policy that operations of the plant be so developed so as to not adversely affect the environment of the surrounding area during emergency situations as well as normal operating periods.

The preservation of life is considered to be of prime importance so that all procedures must be carried out in such a manner as to minimize risk to emergency personnel. Rescue and medical activities are to have priority over all other actions.

Legal Authority:

This plan has been developed so that it complies with the spirit and letter of all applicable federal and state laws pertaining to emergency situations.

*** CITE LAWS ***

1.3   SUBJECT: Situation and Assumptions

Assumptions:

Certain assumptions have been made in the course of the design of this plan, namely:

- that the TOWNNAME fire department, police department, Office of Emergency Services, and other public emergency response organizations will be contacted and will be available to respond to an emergency occurrence and will provide necessary support.

- that COMPANYNAME employees recognize and will carry out their basic responsibilities in an emergency.

- that the local Fire and police departments and other local authorities will assume their responsibility for off-site emergency response.

Situations:

Situations for which the provisions of this plan are designed are those emergency incidents where there is a potential for severe consequences. This includes, but may not be limited to the following situations that would affect the COMPANYNAME and would involve a risk to life, health, the environment, or to property:

- *Technological Disasters*; including a hazardous materials accident, fire, explosion, dam failure, utility failure, etc.

- *Natural Disasters*; including earthquake, flood, windstorm, etc.

- *Health Disasters*; including epidemic, pollution, etc.

- *Social Emergencies*; including bomb threat, arson, riot, labor strife, terrorism, hostage incident, etc.

- *International Crises*; including war, terrorism, etc.

Plant Hazards:

COMPANYNAME has two hazardous materials stored on-site at its plant which might impact the surrounding public. These are

List chemicals and describe how they are used at the plant

Surrounding Area:

Should an emergency occur at the plant, the immediate surrounding area consists of the following land uses:

North:

South:

East:

West:

1.4   SUBJECT: Concept of Emergency Operations

The basic concept of this plan is to provide a comprehensive approach for managing emergencies. The four elements of this approach are prevention, preparedness, response, and recovery.

- *Prevention* incorporates all those activities which eliminate or reduce the probability of a disaster occurring on-site.

- *Preparedness* includes all activities necessary to ensure a high degree of readiness so that response to an incident would be swift and effective.

- *Response* activities are all those measures taken during an incident which minimize damage to the plant and surrounding areas and prevent the loss of life.

- *Recovery* contains those short- and long-term activities which return all systems to normal operations.

Primary responsibility for emergency response activities at this facility have been assigned to COMPANYNAME personnel with the local response agencies agreeing to act in a support role. The fire department will assume control of emergency response activities if and when it deems it necessary.

The authority for responding to minor emergency situations has been assigned to the lowest levels of the response organization possible. If an emergency appears to be localized and does not appear to have the have the potential to become a major emergency, key department personnel in the area of the emergency have responsibility for responding. If the incident appears to be escalating or has the potential to escalate, the highest ranking on-scene person shall contact the control room by telephone, radio, messenger, or other means.

Upon notification of a major emergency or potential thereof, the shift supervisor shall activate begin implementation of response procedures as outline in section 4.0 of this plan. This will result in the activation of the emergency control group (ECG) [defined in Section 1.5] which will oversee and direct response activities in accordance with the provisions of this plan.

1.5   SUBJECT: Organization and Assignment of Responsibilities

Organization:

Responsibility for emergency operations shall be vested in an emergency control group (ECG). This group is assembled immediately and appraises the situation while ensuring that adequate emergency response procedures are implemented. This group shall also establish communications with outside agencies. The emergency control group will augment and support the efforts of other representatives of management.

The emergency control group consists of the general manager, the deputy general manager, the engineering manager, the operations manager, and the safety and loss control specialist. Alternates for each of these individuals shall be designated. Their responsibilities are outlined below.

An emergency brigade is responsible for the implementation of all response activities and operations during an incident. The emergency brigade is staffed for each shift by operating personnel. The responsibilities of the emergency brigade are also outlined below:

Responsibilities:

General Manager:

The COMPANYNAME general manager has final authority for the implementation of this plan. Responsibilities include:

•   Assuring continued compliance with the provisions of COMPANYNAME policy on emergency prevention and response.

•   Approving this plan's provisions and all subsequent revisions.

•   Reviewing and approving the release of any information to the press or public.

•   Approving the initiation of formal or informal agreements with any governmental, community, or industry group.

- Assuring that adequate resources are available to support emergency management activities.

- Monitoring the effectiveness of response activities during emergencies and taking action to ensure that all appropriate procedures are followed.

- Determining the termination of an emergency ("All Clear").

Deputy General Manager:

The deputy general manager will be responsible for fulfilling the duties of the general manager , if the he/she is unable to perform these functions.

In addition,the deputy general manager will act as liaison to local governmental response organizations and the media during emergencies and will participate in public information and community awareness activities.

Engineering Manager:

The engineering manager will be responsible for providing technical advice and assistance to the general manager during emergencies. He will also be responsible for fulfilling the duties of the general manager , if the the general Manager or deputy general manager are unable to perform these functions.

Operations Department Manager:

The operations manager will be responsible for overseeing and directing all on-site response activities during major emergencies. Specific duties include:

- Appointing personnel to the emergency brigade.

- Determining when a full site evacuation is necessary.

- Conducting post-emergency investigations.

Safety Manager:

The safety manager is responsible for developing COMPANYNAME emergency management plans and procedures. Specific responsibilities include:

- Formulating, reviewing, and ensuring implementation of the emergency plan.

- Ensuring that emergency response training concerning the provisions of this plan is provided to all employees.

- Ensuring that this plan is tested through the conduct of drills and exercises.

- Coordinating emergency planning and response activities with local governmental, community or private organizations.

- Administering safety and health review programs.

- Preparing and submitting emergency related reports to management and the COMPANYNAME Board as required.

Emergency Brigade:

The emergency brigade is responsible for the implementation of all response activities and operations during an incident. The emergency brigade is staffed for each shift by operating personnel. The brigade is managed by the following individuals with the following responsibilities.

ShiftSupervisor: The brigade is to be headed by the Shift Supervisor, whose specific duties include:

> • Direction and control of all emergency activities including fire control, spill control, search and rescue, and first aid.

> • Determining a specific plan of action to respond to an incident.

- Assigning brigade members to specific tasks.

- Implementing response procedures as outlined in section 4.0.

Operator III: The Operator III is to assist the Shift Supervisor and assume his/her duties if absent. Specific duties include:

- Assisting the shift supervisor by obtaining personnel and material support as required.

- Establishing and maintaining communications between the Computer Room, the shift supervisor, and government response agencies.

- Recording emergency response events in the operating log.

- Ensuring that other brigade members report to their duty stations as required.

- Receiving and recording headcount information from supervisors and relaying information to the shift supervisor when personnel are missing.

In addition to the above positions, additional personnel may be assigned emergency-related responsibilities as requested by the shift supervisor or as otherwise specified in other sections of this plan.

1.6   SUBJECT: Plan Maintenance and Distribution

It is the responsibility of the safety manager to ensure that all sections of the plan are kept current and are being implemented. The plan is to be reviewed annually to ensure the accuracy of information. The major provisions are to be tested at least yearly through exercises and drills.

The safety manager is to ensure that the plan is modified as required to reflect any changes in the facility's organization or operating conditions.

The safety manager is to ensure that all necessary revisions identified by post incident reviews and investigations are implemented. All future changes and modifications should be forwarded to all possessors of this plan. Distribution of the plan shall be determined by the safety manager but shall at a minimum include the:

Internal Distribution:

- general manager
- deputy general manager
- engineering manager
- operations manager
- safety manager
- maintenance manager
- emergency brigade members
- COMPANYNAME department managers

External Distribution:

- local civil defense agency
- local police department
- local fire department

2.0   PREVENTION

2.1   SUBJECT: General Prevention Policy

It is the primary goal at COMPANYNAME, to provide employees with a safe and healthful workplace. Safety and the prevention of accidents must be an integral part of every task and be given the same attention, effort, and importance that is given to product quality, employee morale, cost, and production.

Accident prevention can be accomplished through risk reduction. Potential hazards must be recognized, evaluated, and controlled so that no unreasonable risks exist.

Responsibility:

Management personnel have the responsibility to see that practices and processes are so engineered, constructed, maintained and operated to provide the utmost in safe and healthful conditions. Line management is directly responsible for ensuring the safety of its employees and the prevention of accidents.

The Safety and Environmental Departments are charged with supporting this effort and providing guidance, consultation, and systems to assist line management in discharging this responsibility.

Employees are responsible for following recognized safety rules, practices, and procedures. Employees are encouraged to detect hazards and inform their supervisors of these conditions and/or unsafe practices.

2.2   SUBJECT: Fire Prevention Policy

Policy:

Fire prevention can be achieved through employee education and adequate safety procedures dealing with flammables and combustibles. It is the policy of the COMPANYNAME to encouraged the active participation of employees in fire prevention programs.

Responsibilities:

The maintenance manager is responsible for ensuring that the provisions of this procedure are implemented within the operations Department. The maintenance manager will incorporate fire prevention and control procedures into the existing safety training program.

The safety manager is responsible for providing overall guidance and direction in the area of fire prevention.

2.3   SUBJECT: Safety and Health Reviews

Purpose:

The purpose of this section is to ensure that all factors and interactions within a new or existing operations that have a potential safety or health effect, are identified, evaluated, and addressed.

Responsibility:

The safety manager will work with engineering and plant operating personnel in reviewing all plant operations before they are implemented to identify potential unsafe conditions and/or potential problems which may lead to health or safety exposures.

Plant personnel shall work with the safety manager to identify potential problems and to identify proper operational procedures. Of primary concern are the operational areas of the plant. Actions to be taken include equipment or procedural changes, development of exposure monitoring strategies, and inclusion of warning statements in procedures.

Procedure:

Each operation shall be reviewed as appropriate or as significant changes occur. Approval of new operating procedures shall be indicated by a stamp on the front page of each procedure signed and dated by the safety and loss control specialist.

A safety and health review will also be applied to the proposed installation and modification of buildings, equipment, mechanical and electric systems, utilities, fire protection systems, grounds, etc.., of a capital nature.

Plans and/or specifications on designated projects shall be submitted to the safety manager for review prior to project implementation.Recommendations will be submitted with the final plans and/or specifications to the applicable department manager for review. If the department manager finds that the final plans and/or specifications do not meet the recommendations of the safety manager, he shall return the final plans and/or specifications to the originator for modifications or a justification of deviations.

2.4   SUBJECT: Inspections

Purpose:

This procedure provides for the inspection and correction of unsafe physical conditions, poor housekeeping, and a spot check for unsafe practices.

Responsibility:

Each area supervisor/manager is responsible for ensuring the timely completion of periodic inspections and correction of problems. The actual inspection may be delegated to a subordinate.

Frequency:

The frequency of specific inspections is as follows:

        Plant   - Monthly
        Emergency Equipment   - Weekly
        Lab   - Monthly
        Maintenance   - Monthly
        Office   - Quarterly
        Fire Protection Systems   - Monthly, Annually

The safety manager will conduct a periodic inspection with each responsible department manager in their area of responsibility.

2.5   SUBJECT: Plant Safety Committee

Purpose:

The purpose of the plant safety committee is to encourage teamwork in addressing safety concerns and promoting safety ideas.

Personnel assigned to the safety committee are listed in attachment 6.7.

Procedure:

The plant safety committee will meet monthly for approximately one hour. Ideas and concerns presented at the meeting should be discussed, agreed upon and corrective actions taken. If a corrective action requires a maintenance work request, it should be written immediately.

The plant safety committee secretary will keep minutes of each meeting. The minutes will include the names of those present, items and concerns presented at the meeting, action taken, and any items not resolved. Minutes will be posted at various COMPANYNAME locations.

3.0   PREPAREDNESS

3.1   SUBJECT: Training

Purpose:

Training programs designed to ensure the continued competence in proper emergency response skills and in the procedures established by this plan are to be conducted on a continuing basis as outlined in the COMPANYNAME safety directives.

Responsibility:

Development and implementation of emergency response training is the responsibility of the safety manager with the director of the plant safety committee and with the assistance of the Maintenance Safety Supervisor and the departments.

Curriculum Overview:

Training curriculum and course content are to be based upon the task requirements and special hazards associated with the emergency situations. Basic requirements for training various individuals for emergencies include the following:

> • All Employees   - Evacuation Procedures
> - Reporting Emergencies
> - Hazardous Materials Safe
>   Handling
> - Personal Protective Equipment
> - Fire Safety
> - Use of Fire Extinguishers
> - First Aid/CPR
>
> • Supervisors   & Managers   - Same as above
> - Plan Provisions
> - Risk Assessment/Loss Control

- • Brigade Members   - Same as above
                                     - Use of SCBA
                                     - First Aid
                                     - Search and Rescue Techniques
                                     - Use of Communications
                                         Equipment
                                     - Fire Protection System
                                     - Spill Control

Where practical, emergency response training will be incorporated into existing safety and operational training. Training of skills is to employ hands-on, practical drills.

Training programs will include testing of student proficiency where the level of expertise requires demonstrable skills. All training and testing is to be documented for each employee by the safety and loss control specialist. Emergency training is to be repeated at a frequency stated on the schedule contained in attachment 6.8.

Joint Training:

Joint training sessions of COMPANYNAME and off-site response personnel (both government and private organizations) are to be conducted annually. This training is to include site orientation tours for off-site personnel.

Training Program Evaluation:

Training content should be reviewed annually and modified as necessary to ensure that training adequately reflects changes in hazards and conditions.

3.2   SUBJECT: Drills and Exercises

Purpose:

While drills and exercises can be used for training purposes, their primary function for this plan is to provide the means of testing the adequacy of the plan's provisions and the level of readiness of response personnel.

Responsibility:

The safety manager is responsible for coordinating the development of and assisting in conducting drills and exercises.

Types of Exercises:

The following types of drills and exercises are to be used:

- *Tabletop exercises* involve presenting to key emergency personnel a simulated emergency situation in an informal setting to elicit constructive discussion as the participants examine and resolve problems based on the plan.

- *Functional drills* are practical exercises designed to test the capability of personnel to perform a specific function (i.e., communications, first aid, rescue).

- *Full-scale exercises* are intended to evaluate the operational capability of the COMPANYNAME's emergency organization and the adequacy of this plan.

Frequency:

Tabletop exercises are to be conducted after initial implementation of this plan and after any major revisions of this plan or changes in key personnel.

Functional drills for various emergency functions are to be conducted at least annually. Evacuation drills should be conducted annually. Emergency brigade drills are to be conducted quarterly on various types of response activities (i.e. chlorine response, first aid, etc...).

A full-scale exercise is to be conducted annually involving all emergency response personnel as well as off-site organizations.

Preparations:

Preparation for a drill or exercise will vary depending on the type and scope involved. However, preparation and planning should include:

- Plan review and identification of possible problem areas.

- Establishing objectives.

- Identifying resources to be involved including personnel.

- Developing exercise scenarios, a major sequence of events list, and expected action checklists.

- Assigning and training controllers and evaluators.

The scenario used will be realistic and based upon current operating conditions. The primary event (fire, spill, etc...) is to be determined based on the objective of the exercise.

A sequence of major events list is to be developed to help stimulate an actual emergency incident. Expected responses for each major event are to be determined. Conditions are to simulate, as closely as possible, actual emergency situations.

Follow-up:

Results of drills and exercises are to be reviewed by the participants, evaluators, and the safety manager to identify problem areas such as deficiencies in the plan, training, personnel or equipment.

The safety manager is to prepare a final report and submit it to the plant safety committee for implementation of corrective actions.

3.3   SUBJECT: Facilities, Supplies, and Equipment

Purpose:

To ensure an effective response to emergency situations, adequate quantities and types of supplies and equipment are to be maintained on-site for use by the emergency brigade and others.

Responsibility:

Responsibility for maintaining Plant Operating Division equipment in a ready state and for determining the adequacy of equipment is assigned to the maintenance manager.

Procedure:

An inspection of all emergency equipment is to be performed monthly by the maintenance manager in accordance with the attached schedule. Records of this inspection will be kept on file in the maintenance manager's office.

Life support equipment (e.g., self contained breathing apparatus) is to be tested in accordance with manufacturer's instructions to ensure its reliability. Records of all tests are to be maintained by the maintenance manager.

Provisions for sanitizing such items as respirators shall be made available by the maintenance manager. Where fit and sanitation are of concern, equipment should be preassigned and labeled.

Items with a limited shelf life or items such as sterile first-aid supplies should be rotated where possible and replaced as required.

All Plant Operating Division items expended during an emergency or exercise, as well as damaged, defective or spoiled items, are to be reported to the maintenance manager who is to arrange for immediate replacement.

A list of vendors capable of providing the immediate emergency resupply of items expended during sustained operations is to be maintained by the COMPANYNAME Purchasing Department.

3.4   SUBJECT: Mutual Aid

It is the policy of COMPANYNAME to assist its neighbors in preparing for and responding to emergencies that may occur either on-site or off-site. COMPANYNAME provides response assistance to local emergency response agencies and to neighboring industrial facilities.

COMPANYNAME's assistance varies depending on the unique circumstances of each incident but may include technical advice or provision of COMPANYNAME resources. The Resource Manual is located in the safety manager's office.

COMPANYNAME provides this assistance under verbal agreement with local officials. Assistance is only provided in cases where plant property and personnel would be reasonably safe from harm.

3.5   SUBJECT: Public Information

Purpose:

The purpose of this procedure is to provide for a program that ensures the distribution of information concerning COMPANYNAME emergency management programs to the media and the general public.

Objectives:

The objectives of the COMPANYNAME public information program are:

- To help foster the view that COMPANYNAME is responsible on issues concerning environmental, health and safety.

- To clearly communicate COMPANYNAME policies, positions, and activities to the public.

- To inform the public about the positive contributions of COMPANYNAME to the community.

- To monitor and interpret the agendas of key community interest groups for COMPANYNAME management.

- To produce materials as needed to further internal or external understanding of key issues.

- To encourage collaboration among business, government, and citizen groups within the community.

Responsibility:

The deputy general manager is responsible for developing, coordinating, and implementing all elements of the COMPANYNAME Community Awareness Program. The safety manager will provide assistance as necessary.

To ensure effectiveness, management will participate in this program to the extent possible. The general manager and other managers are expected to participate in public speaking engagements.

Procedure:

No COMPANYNAME employee other than the general manager, or the deputy general manager shall respond to questions from the media. All media contacts should be referred to one of these individuals.

The deputy general manager shall develop a list of those members of the COMPANYNAME staff who might be expected to participate in the Public Information Program by providing tours, public speaking engagements or dealing with the press. These individuals should be provided specialized training, handouts, etc., concerning public relations.

The following community awareness programs will be developed and utilized to achieve the above stated objectives:

A **slide program** is to be presented to various community groups. The slide show will contain the following information:

- Description of COMPANYNAME
- History
- Location
- Description
- Safety Record
- Environmental Protection Programs
- Contributions to the Community
- Jobs provided
- Local dollars expended
- Overview of Emergency Plan
- Coordination with off-site authorities
- Joint training, drills, and exercises
- Mutual aid assistance

The deputy general manager will identify community organizations to whom this program should be offered. Among the groups to be considered are:

- Local Schools
- United Way
- American Legion
- Elks Lodge
- Jaycees
- Masonic Lodge
- Rotary Clubs
- VFW
- Knights of Columbus
- League of Women Voters
- Local & State Government Response Agencies

The presentation of this slide show however is not restricted to the above organizations and any group (i.e., professional organizations, community volunteer groups, etc.) requesting the presentation will be given due consideration.

It is expected that some groups will express more interest than others in receiving this presentation. Letters explaining the availability of this program should be sent to each of the above groups.

A **pamphlet** which contains information similar to the slide show will be distributed during public meetings. The pamphlet will also be given to neighboring industrial facilities in quantities sufficient for distribution to all employees. This pamphlet should also be distributed in conjunction with the presentation of the slide show. Other groups will be provided with copies of the pamphlet upon request.

Following distribution to neighboring facilities, additional distribution should only take place every three years. Distribution shall be determined by the deputy general manager.

A **media information kit** will be developed by the Public Information Coordinator and distributed to members of the press during interviews, public meetings, and emergency incidents. The kit will include an expanded version of the information presented in the slide show and pamphlet. It will also contain current policy statements, plant safety record data, and other news worthy information. A supply of 25 media information kits will be on hand at all times and shall be updated quarterly by the deputy general manager or designee.

On-site visits and tours shall be provided to interested groups which will allow for a first hand look at COMPANYNAME operations. The program shall consist of formal presentations and a tour of the plant as indicated below:

- Introductions - deputy general manager
- COMPANYNAME and its Policies - general manager
- Safety, Health, and Emergency Management - safety manager
- Tour of the Plant - operations manager

This program may be offered to the following groups:

- Local government officials
- Representative of the local media
- Management representatives of nearby companies
- Other groups as determined by the deputy General Manager

This program shall be offered at least every third year.

COMPANYNAME recognizes that its employees are the best and most effective communicators of COMPANYNAME programs and policies. Therefore, the deputy general manager will annually provide an update of the plant's community awareness and emergency management programs to all employees.

The deputy general manager shall keep a record of all community awareness activities in which COMPANYNAME and/or its employees participate. A summary of these activities shall be annually forwarded to the Board of Directors. It shall also be made part of the media information kit.

Program Maintenance:

The deputy general manager is responsible for maintaining and updating this procedure. Changes to this procedure shall be coordinated with the safety manager and approved by the general manager.

4.0   RESPONSE

4.1   SUBJECT: Notification, Direction and Control

Purpose:

The purpose of this procedure is to outline provisions for the direction and control of emergency operations at COMPANYNAME. It specifies who is responsible for overall emergency management as well as logistical aspects of emergency operations. The procedure also specifies notification protocols.

Responsibility:

The general manager is ultimately responsible for managing all emergency situations at the plant. The operations manager has been vested with the responsibility and authority for all emergency response operations within the plant. The operations manager has authority to utilize all personnel and plant resources necessary to contain and control emergency incidents at the plant.

Primary Control Center:

The plant control room and the adjacent administrative offices are activated to allow for central control of all activities necessary to support emergency operations. The Emergency Control Center consists of the plant control room (for operations staff), the operations manager's office (for senior managers), and the large conference room (for media briefings).

The plant control room is stocked with the following: radio equipment, telephones, office equipment and supplies, area maps (showing locations of commercial and industrial plants, residences, and roads within four miles of the treatment plant), emergency plans, technical manuals, plant blueprints and maps, SCBA, protective clothing, environmental testing equipment, and furniture.

Alternate Control Center:

The alternate control center is located in the warehouse control room. The alternate emergency control center is similarly stocked. This alternate control center is to be activated in the event that the primary center (plant control room) becomes unusable.

Emergency Activation:

The Emergency Control Center is activated when the shift supervisor determines that a major emergency is underway or probable. The shift supervisor or designee will notify the following personnel:

- general manager
- deputy general manager
- engineering manager
- operations manager
- safety manager
- Local Emergency Response Agencies, as required [see attachment 6.1]

When an emergency occurs, members of the emergency brigade will take appropriate actions and await instructions from the shift supervisor. [If the emergency involves chlorine, they will immediately implement chlorine emergency procedures.]

Upon activation, the shift supervisor is responsible for the following:

- Receive initial assessment of incident and determine need for evacuation.

- Ensure that all personnel are accounted for. Each evacuation station should report the headcount to the coordinator. The shift supervisor will instruct the emergency brigade to search for missing individuals.

- Ensure that outside agencies are notified as necessary (i.e., fire department, Ambulance, etc...). The attached Notification List [attachment 6.1], should be utilized in contacting necessary government agencies.

The fire and police departments will pull their emergency vehicles up to the main gate. The local fire department's on-scene commander will take over command of emergency operations if he determines that the plant emergency brigade is unable to handle the situation. Requests for additional community response assistance will be requested by the general manager through the fire department's On-scene Commander.

4.2   SUBJECT: Emergency Communications

Purpose:

The purpose of this section is to describe the communications methods to be used during an emergency.

Internal Communications:

The primary means of communications between the shift supervisor and the emergency brigade will be two-way radio sets. The COMPANYNAME emergency radio system includes two base stations located at the Control Room and the safety department. During an emergency, portable radio units shall be provided to emergency brigade members and/or strategically placed at various locations as directed by the affected department manager or shift supervisor. The operation of this system during emergencies is determined by the emergency control group which will monitor inter-plant communications to the extent necessary.

The Maintenance Manager is responsible for maintaining plant owned radio equipment. The emergency radio communications system will be tested weekly to ensure a condition of readiness at all times.

An alternative communications medium in the plant is the telephone. This system is used on a daily basis and all employees are familiar with its proper functioning.

Employees are instructed to use the telephone only for essential calls during an emergency.

External Communications:

The method of communicating with off-site authorities is through the use of the telephone. If telephone lines are out of order and initial notification has not been made, the shift supervisor will send an employee messenger to the nearest fire station.

4.3   SUBJECT: Emergency Public Information

Purpose:

This procedure provides for the managed release of information to the public during and following an emergency situation.

Policy:

In order to avoid the release of confusing, contradictory, or misleading information, only the general manager, or the deputy general manager are authorized to speak to the media on behalf of COMPANYNAME, without the written authorization of the general Manager. All persons authorized for this activity will receive proper training in dealing with the media.

Responsibility:

The deputy general manager, is responsible for preparing press releases and other materials for release to the media during an emergency.

Press Center:

During an emergency situation, access to the plant is to be denied to the press and public. Security Guards and COMPANYNAME personnel are instructed to direct the press and public to the Administration Building's large conference room entrance (located at the front of the Administration Building) during normal business hours.

The press center is to be established by the deputy general manager in the large conference room of the Administration Building. Two employees are to be assigned as monitors to ensure that access is confined to this area. These employees shall not discuss the emergency incident or plant operations with the media. Management will provide periodic briefings.

At night, if no COMPANYNAME personnel are available to monitor the media inside the large conference room, the media is to be kept outside of buildings and fenced areas but away from the gate so as to allow access for fire department vehicles.

Press Releases:

The general manager must approve all press releases prior to their dissemination. The statement should outline:

- The nature and extent of the emergency incident.
- Response actions underway.
- Impact on off-site areas.
- Coordination with off-site officials.

Media Guidelines:

In dealing with the press, the following guidelines should be adhered to:

- Regularly scheduled press releases should be issued. If a statement is promised...it should be delivered!

- Only accurate, substantiated information is to be released. Do not speculate. Do not attempt to place blame. Do not mislead.

- If a request for information is to be denied, explain the reason(s) for denial.

- Be sensitive to the rights of the media and public to know how the incident will affect the community and the environment.

- Plan regular follow-up releases and statements after the emergency. Consider inviting media representatives and others to visit the emergency site, when safe to do so.

- Do not release estimates of damage nor allow photographs on site without first consulting the general manager.

- All press inquiries and interviews should be logged for future reference.

Rumor Control:

News stories should be monitored to ensure that only factual information is being distributed by the media. Any misinformation or rumors should be quickly corrected or dispelled by the deputy general manager.

4.4   SUBJECT: Evacuation and Personnel Accountability

Purpose:

In the event of an emergency with the potential to threaten the life and safety of employees, an evacuation of the area may be necessary. This procedure outlines provisions for evacuating treatment plant facilities.

Emergencies occurring in the plant area may require the evacuation of plant personnel, but may not involve the evacuation of the Headquarters Office Building. Emergencies occurring in the Headquarters Office Building may not require the evacuation of plant area. The operations manager or shift supervisor will determine if the evacuation of other areas is necessary.

PLANT EVACUATION PROCEDURES:

If necessitated by a chlorine emergency, fire, or sulfur dioxide emergency, evacuation procedures for plant personnel are as follows:

- The shift supervisor will determine whether or not an evacuation is necessary. If an evacuation is warranted, the shift supervisor will:

    • Sound the siren (15-second duration)
    • Make Public Address Statement
    • Sound the siren a second time

- Upon notification of an evacuation, employees should exit the building, check wind direction by looking at windsocks located throughout the plant, and proceed to the nearest evacuation station upwind of the hazard (see attached map). Appendix 6.3 contains a map of the plant and indicates the location of wind socks and evacuation stations

- Area supervisors are responsible for checking washrooms, etc., to ensure that all employees have evacuated the area.

- COMPANYNAME employees will escort visitors to the appropriate evacuation station.

PLANT EVACUATION HEADCOUNT PROCEDURES:

To ensure that all employees have safely evacuated, the following headcount procedures will be followed:

- All employees will report to one of three on-site evacuation stations. If an off-site evacuation is necessary, then employees will be instructed to report to a designated location.

- The first supervisor arriving at an evacuation station is responsible for conducting the headcount. Each employee will orderly call out his/her identification number and the supervisor will check off those reporting in on a roster. Upon completion of the headcount, the supervisor will call the Emergency Control Room on extension 911 and specify who has checked in.

- Any supervisor who is aware of an employee who is out on sick leave or away from the plant for any reason, will report this to the headcount supervisor. This information will be transmitted to the plant control room as well.

- The shift supervisor will order the emergency brigade to initiate a search for any missing employee or visitor upon completion of the headcount.

ADMINISTRATION BUILDING EVACUATION PROCEDURES:

Evacuation procedures for the Administration Building are as follows:

- When the alarm sounds and instructions are given, personnel will turn off all non-essential equipment and proceed to the first floor lobby using the west or east stairs. Employees should accompany to this area.

- Personnel should assemble at the following locations as appropriate:

  • Personnel who work on the first floor should assemble in the visitor waiting area.
  • Personnel who work on the second floor should assemble at the permit counter.
  • Personnel who work on the third floor should assemble at the

receptionist's counter.

- Area supervisors will check washrooms, etc.., to ensure that all personnel have left and should check to ensure that no security doors are left unlocked to facilitate access. .

- Escape respirators will be distributed at the assembly stations in the event of a chlorine emergency.

ADMINISTRATION BUILDING HEADCOUNT PROCEDURES:

To ensure that all employees have safely evacuated the Administration Building, the following headcount procedures will be followed by both plant and office areas:

- All employees will remain in the assembly area until instructed to do otherwise.

- All employees will be instructed to remain quiet and to form groups by department to facilitate accounting. Group Leaders will take roll call to determine that all employees are accounted for.

- Group leaders will notify the Receptionist of any missing individuals. The group leader will also notify the receptionist of any visitor who may be in the assembly area.

- The receptionist will check the visitor's log for names of visitors who might be in the building and compare the log to the group leaders report. The receptionist will report the names of any missing employees or visitors to the plant control room by dialing extension 911.

- If an off-site evacuation is deemed necessary, the Plant shift supervisor will announce the information over the public address system.

The operations manager or in his absence, the shift supervisor, may order an evacuation of personnel off-site if conditions warrant. All non-essential personnel will be instructed via the Public Address System to leave the facility and proceed to an off-site location.

4.5   SUBJECT: Government/Private Agency Coordination

Purpose:

The extent of involvement, if any, by government agencies (ie. fire, police, etc.) and/or private organizations (ie. hospitals, telephone company, etc.) in aplant emergency, will depend upon the type and magnitude of the crisis. This section outlines the types of assistance available from outside organizations during treatment plant emergencies.

Outside agencies which may be utilized in the event of a major emergency are as follows [telephone numbers are contained in attachment 6.2]:

Organization:	Fire	Spill	Security	Medical	Tech Assist	Other
Fire department	√	√		√	√	
Police department			√	√		
State Police			√		√	
US Coast Guard			√		√	
Local Civil Defense					√	√
State Civil Defense					√	√
CHEMTREC					√	
American Red Cross						√
Local Hospital				√		

Special Provisions:

Law Enforcement Agencies - must be relied upon to maintain order in area adjacent to the plant and to expedite the movement of service vehicles (fire trucks, doctors' vehicles, ambulances) into the plant. These agencies will also coordinate community evacuation should this become necessary.

Coast Guard - will coordinate river traffic when emergency conditions involve waterways and will augment fire/spill control measures in waterway incidents.

Railroads - should be contacted to preclude any unnecessary rail traffic from hindering emergency vehicles attempting to enter the area.

Civil Defense (state and local) - will coordinate the evacuation of the community, provide emergency aid to the injured, aid in information dissemination, and assist in rescue operations.

Red Cross - will provide shelter for the evacuated, and supply necessary personal comfort items and food.

4.6   SUBJECT: Fire/Explosion

Purpose:

This section outlines the provisions for responding to a fire and/or explosion in the plant.

Plant Area - Employee Responsibility:

In the event that an employee discovers a fire or explosion, he/she should:

- Attempt to extinguish a small fire, but only if there is backup support and only if he/she has been trained in the proper use of the extinguisher.

- If the attempt fails to extinguish the fire immediately, the employee should immediately notify the shift supervisor via one of the following methods:

  - Telephone Extension 111
  - Plant radio system (222)
  - Fire alarm pull box (place follow-up call to Shift Supervisor as soon as possible)

- Provide the following information:

  - Exact location of fire
  - Type of fire (i.e., electrical, flammable liquid, or combustible material)
  - Whether or not the fire is near a critical system (i.e., major equipment, chlorine system, fuel gas lines, steam lines, etc.)
  - Whether or not medical assistance is needed.

If a fire is announced, all other employees should follow the instructions of the shift supervisor.

Shift Supervisor Responsibility:

When advised of a fire/explosion, the shift supervisor will immediately ascertain the nature, extent, and location of the fire/explosion and formulate response tactics.

- The shift supervisor will direct the emergency brigade to proceed to the impacted area and begin response activities.

- The emergency brigade is to advise the shift supervisor of the situation and request any additional assistance or equipment as necessary.

Upon notification of a fire/explosion, the shift supervisor will notify the local fire department by dialing 9-1-1. As necessary, the shift supervisor will initiate evacuation procedures. During off hours, the shift supervisor will contact appropriate management personnel utilizing the Emergency Notification List (Attachment 6.2).

The shift supervisor is to respond by directing a person to proceed to the front gate to act as a liaison between plant personnel and the local Fire and police departments.

Designated management and technical support personnel (reference Section 4.1) are to report to the Administrative Offices and provide support as required.

Office Lab Area:

Procedures for responding to a fire in the office or lab areas are generally the same except:

- Initial notification of an emergency by an employee in the office/lab is to be through the use of the "pull box" stations.

- The plant emergency brigade is to be summoned by means of the PA system.

  [NOTE: Plant personnel need not evacuate for fires in the office or lab areas unless ordered by the shift supervisor.]

Emergency Brigade:

Specific fire extinguishing tactics to be used by the emergency brigade are to be determined on a case-by-case basis; however, the following general principals apply:

• No actions are to be taken that will subject the brigade personnel to unreasonable risk.

• Rescue and first aid take priority over fire suppression.

• The possibility of explosion must always be considered.

• Sprinkler systems are not to be shut off except by order of the brigade Chief after all evidence of fire is gone.

4.7 SUBJECT: Hazardous Material Spill

Purpose:

The purpose of this procedure is to minimize the safety, health and environmental hazards due to spills of hazardous materials. Hazardous materials located on COMPANYNAME property include chlorine and sulfur dioxide. The location of these hazardous materials are indicated on the map in attachment 6.5.

For the purposes of this plan, a spill is defined as the accidental release of a liquid or solid materials from its proper container whether from container failure, upset, or unintentional drainage.

Small Spills:

Small spills which pose no safety and health dangers and are not likely to adversely affect the environment, are to be handled by trained area personnel with appropriate materials on hand. These personnel should:

- Eliminate the source of the spill by closing valves, righting drums, turning leaking drums over, etc...

- Prevent the chemical from entering the drainage system.

- Add neutralizing agents and/or absorbents (a list of these materials and their location is included in attachment 6.6).

Large Spills:

For large spills or those that present a threat to health or safety due to toxic fumes, flammability, or the possibility of release into the environment, the emergency brigade is to be notified by using the siren and PA system to announce the nature of the emergency.

The emergency brigade is to respond as in a fire but with appropriate spill control equipment and initiate containment procedures.

- Eliminate the source of the spill by closing valves, patching leaks, righting containers, emptying faulty containers, etc...

- Prevent or contain the extent of the spill through diking or other means.

- Implement other spill-control measures as appropriate for the spill

Evacuation:

The shift supervisor should commence evacuation (refer to Evacuation Procedure) of the area if the following type of chlorine or sulfur dioxide leak has occurred:

- A major leak (50'x 50'x 50' cube).
- A minor but continuing dangerous leak.

If the shift supervisor judges that a health and safety hazard exists, he/she may declare a full-site evacuation of all nonessential personnel.

The local fire and police departments, the local civil defense agency, and the local police department shall be notified by the operations manager or his designated representative whenever a chlorine or sulfur dioxide leak cannot be contained or an evacuation has taken place.

The safety manager is responsible for contacting all federal and state agencies requiring notification.

4.8   **SUBJECT: Sabotage**

Purpose:

This section is intended to provide an effective means of preventing acts of sabotage by employees of COMPANYNAME or others and of reacting to such acts.

This section defines sabotage as any deliberate action by an individual or group designed to cause harm to COMPANYNAME personnel, damage equipment, or disrupt normal operations.

Critical Equipment:

The safety manager will construct a list of critical pieces of production equipment and stores of materials (that by reason of easy access, isolated location, or that require only common knowledge to operate) that are considered susceptible to sabotage. For those items identified, the following procedures should be implemented.

- Access is restricted.

- Devices are locked where possible.

- Tamper or intrusion alarms are used.

Facility Access:

Access to COMPANYNAME facilities by vendors, contractors and other visitors is controlled in accordance with the COMPANYNAME Gate Control Procedure.

- All visitors are logged in and out.

- Visitors must wear a special ID badge.

- Visitors are escorted by COMPANYNAME employees.

- Contractors, vendors, delivery drivers, etc., are to be restricted to the immediate location of the work area.

- Remote entrance points are to be kept secured except as needed.

Other Provisions:

The Administrative Office Manager is responsible for developing and maintaining access cards and key control procedures.

The Personnel Department should immediately inform appropriate supervisors of all terminations.

All managers and supervisors should immediately report suspected acts of sabotage to the safety manager and Operations Manager.

Since the results of an act of sabotage are the same as those resulting from accidental events (i.e., fire, explosion, spill, etc.,), appropriate response procedures are to be used as appropriate.

Specific individuals are to be assigned responsibility for securing designated doors, gates, cabinets, etc., at the end of shift periods.

All issues relating to sabotage or other security questions should be referred to the local police.

4.9   SUBJECT: Bomb Threat

Purpose:

The purpose of this procedure is to provide for the proper handling of a bomb threat received at COMPANYNAME.

Telephone Threat:

When a bomb threat is received by telephone, the person receiving the call is to attempt to gain as much information as possible, including:

* All information about the device itself, including set time, type, description, location.

* Reason for making call (angry with company, extortion, etc.,).

* Any information about the caller (apparent age, voice characteristics, speech, language, accent, manner, use of unusual terms).

* Any information of the location of the caller (inside or outside a building, background noises, etc.,).

The person receiving the threat should then contact the general manager or deputy general manager or Operations Manager.

General Manager's Responsibility:

The general manager will contact the police department immediately. The police will advise COMPANYNAME on the next course of action. Possible actions may include:

• Inform department heads but do not search or evacuate.

• Initiate search but do not evacuate.

• Evacuate specific area and search.

• Initiate emergency shutdown procedures and end shift early.

• Search and then full evacuation immediately prior to target time.

• Immediate evacuation.

Factors to be considered in determining appropriate levels of response include:

• A judgment of sincerity of threat based on information contained in it.

• Any existing labor problems or known disgruntled employees.

• Previous threats received.

• Recent attacks against other facilities in the area.

• Time allotment.

Police:

Searches are to be conducted by police with the assistance of department personnel who are most able to spot "out-of-place" items.

Only Police personnel are to handle a suspected device.

Emergency Brigade:

The emergency brigade is to be on standby to facilitate an immediate response to an actual detonation (refer to fire/explosion section).

4.10   SUBJECT: Natural Hazards

Tornado/High Winds:

If a "tornado watch" is issued by the National Weather Service, personnel should be assigned to monitor weather conditions, listen for broadcast warnings, and report on threatening conditions.

If a "warning" is issued by the National Weather Service (meaning that a tornado has actually been sighted in the area) or if a funnel cloud is seen by plant personnel, the following steps are to be taken:

- Personnel will be notified by the public address system.

- The emergency brigade is placed on alert.

- Plant personnel are to seek shelter in the administration building basement, ground-level interior rooms, plant tunnels, restrooms, or hallways.

- All nonessential utilities should be shut off.

If no advance warning is received, employees are to seek shelter in the areas listed above. If this is not possible, employees should seek safety under a table or heavy piece of equipment which offers protection from falling debris.

After the passing of a tornado, personnel should inspect their areas for damage. If the plant was struck, emergency brigade personnel will begin rescue, first-aid and damage control activities. Damage assessment, cleanup and restoration, and other recovery activities should follow.

Floods:

When warnings of an impending flood conditions are received via weather broadcasts, U.S. Weather Service, or the police/fire department, the following steps should be taken:

- All movable equipment and supplies are to be moved to second floor levels or to other elevated areas.

- Outside areas are to be checked for equipment and materials that could be damaged by flood waters.

- If time allows, sandbagged dikes are to be constructed to augment existing dikes and to protect high risk items.

- Storage tanks/vessels should be checked and secured.

4.11   SUBJECT: Major Earthquakes

Purpose:

The purpose of this section is to outline activities to be taken during and following an earthquake to protect employees and to continue vital operations.

Employee Safety:

During an earthquake, all personnel should evacuate buildings and proceed to areas away from walls, windows, or power lines. If evacuation is not possible, employees are to seek shelter under a desk, table, etc.., or in doorways that offer protection from falling objects. After the initial quake, aftershocks should be expected.

The shift supervisor should contact operators by radio for a report on employee safety and a condition of plant facilities and equipment. The emergency brigade should begin rescue, first-aid, and damage control activities.

Emergency Shutdown:

As soon as possible, emergency shutdown procedures should be implemented, including securing chemical and steam systems:

- Secure liquid and gas valves on chlorine tank car.
- Secure boiler and close main steam stop.
- Secure liquid and gas valves on SO_2 storage tank.

Close main influent gate and assess damage.

Emergency Primary Treatment:

In order to provide primary treatment following an earthquake when utilities are out, the following should be implemented:

- Activate propane gas system.
- Start generators for electrical service.

5.0   RECOVERY

5.1   SUBJECT: Recovery Introduction

The procedures outlined in this section are intended to minimize the adverse effects of an emergency incident by providing an effective means of reestablishing normal operating conditions in the least amount of time. In addition, the procedures also provide for determining the cause of the accident so that future incidents can be prevented. Emergency response activities will be reviewed so that more effective procedures can be used in the future.

This section is divided into five parts:

- Incident Investigation

- Establishing a Recovery Team

- Damage Assessment

- Cleanup and Restoration

- Postemergency & Recovery Reporting

5.2   SUBJECT: Incident Investigation

As soon as it is safe to do so, an incident investigation to determine the cause of an emergency and to determine the means of preventing any further occurrences is to be conducted.

Responsibility:

The operations manager will coordinate the investigation and prepare all final reports for submittal.

The investigation is to be conducted by a team to be headed by the operations division manager and consisting of appropriate personnel including all or some of the following:

- engineering manager

- safety manager

- Shift Superintendent

- Plant Maintenance Superintendent

Procedure:

The investigation team should immediately seal off the incident scene and commence its investigation in order to minimize the tampering with or removal of any physical evidence.

The investigation of the scene should include:

- Photographing the area

- Determining the point of origin of the fire/leak/ explosion, if applicable.

- Note the position of valves, controls, and devices.

- Note any unusual items in the area or any damage that is inconsistent with the type of incident or intensity.

Written or recorded statements from all operators involved, potential witnesses, and others, who might have pertinent knowledge of the incident. Descriptions of events leading up to the incident, assessment of what might have happened, why, and suggested corrective actions should be obtained.

Where possible (and when no risk of another incident exists), tests of equipment and/or simulations should be employed.

Report:

A final report is to be prepared to include the most probable cause(s) and recommended corrective measures. The report should consider:

- Failures of Equipment

- Failures of Maintenance

- Failures of Procedures

- Inadequate Training

- Human Error

Corrective Actions:

The Investigation Team is also responsible for conducting a review of response activities during the emergency to evaluate the adequacy of training, equipment, and procedures.

The operations manager is responsible for ensuring that all corrective actions to prevent a recurrence of the incident and provide better response to emergencies are taken.

5.3   SUBJECT: Recovery Team

Purpose:

In order to facilitate the restoration of normal plant operations as soon as an emergency situation has been declared under control, a Recovery Team is to be established to manage recovery activities, including damage assessment. The Recovery Team is to be activated by the operations manager

Organization:

The Recovery Team's organization will vary depending on the nature of the incident and the extent of recovery operations. As a general rule however, the following individuals should be included:

- The operations division manager to act as team leader.

- The Purchasing and Materials Officer to expedite contracts.

- The engineering manager and/or appropriate plant engineers or technical support personnel (e.g., chemists).

- The Accounting Supervisor to provide assistance in maintaining accurate accounting of losses.

- The Maintenance Manager to ensure proper Maintenance Department support.

- The safety manager to provide liaison with risk management department, prepare final reports and to assess the incident's impact on the environment and to develop necessary government reports.

The recovery team will be responsible for damage assessment, clean-up and salvage operations, and the restoration of operations. The operations manager will forward weekly status reports to the general manager and others as necessary, including all estimates of repair costs, inventories, and schedules.

Each member of the team is to designate an alternate person to act in their stead when necessary. Other departments may be called on by the team leader to assist in recovery operations.

5.4   SUBJECT: Damage Assessment

A primary function of the Recovery Team will be to assess the damage to plant structures, equipment, and materials as well as to assess damage to off-site areas and the environment in conjunction with civil authorities.

Responsibility:

The operations division manager will coordinate the team's efforts and assign tasks and areas of responsibility.

The Plant Maintenance Superintendent will supervise the inspection and testing of equipment by qualified maintenance personnel.

The operations Superintendent will consult with the Plant Maintenance Supervisor and technical personnel, to develop a list of necessary replacement items and services.

The Purchasing and Material Officer is to contact vendors of damaged items to obtain any necessary technical assistance.

The Accounting Supervisor is to ensure that all damaged items, labor costs, etc.., are recorded and that proper documentation is maintained.

Each team member will report their findings to the operations division manager.

The safety manager will follow up on any injured individuals to:

- Ensure that the attending physician is given all appropriate technical support.

- That proper notification of next of kin has been made.

- Obtain the prognosis and estimates of medical costs.

- Determine reporting requirements to OSHA and others.

Based on the findings of the Recovery Team, an action plan will be prepared which includes the following.

- An estimate of losses detailing equipment, machinery, personnel injured, and materials lost.

- A prioritized listing of necessary repair/reconstruction work, personnel assignments, and estimates of completion schedules.

5.5   SUBJECT: Clean-up and Restoration Operations

As soon as incident investigations are completed, and restoration plans have been made, cleanup and restoration activities should commence.

Responsibility:

The safety manager is responsible for the proper reporting and documentation to necessary federal, state and local agencies.

The Maintenance Superintendent will direct repair, cleanup and restoration of utilities and salvage activities.

The operations division manager is responsible for coordinating the activities of contractors/vendors with in-plant personnel (such as the Purchasing Department) and supply all necessary reference information.

The maintenance manager is responsible for monitoring all activities to ensure the safety of cleanup and salvage personnel making sure that proper personal protective equipment is utilized. The maintenance manager should replace or restore all emergency equipment and supplies as necessary.

The purchasing manager will expedite all orders for necessary equipment, supplies, and services.

The accounting supervisor will monitor and record all costs related to recovery activities.

All recovery team members will submit weekly progress reports to the operations division manager. A consolidated report will be forwarded to the general manager.

Damaged Equipment:

Temporary storage facilities will be established for damaged equipment to facilitate inspection by insurance personnel. Prior to restoration of normal operations, all affected equipment is to be tested and checked out according to procedures established by the operations manager.

5.6   SUBJECT: Postemergency Recovery Reports

As soon as is practical after the secession of the recovery phase, the safety manager is to hold review sessions with emergency response personnel to evaluate:

- The adequacy of emergency response procedures.

- The adequacy of the investigation of the cause of the incident.

- Summarize the postemergency activities.

The safety manager will then prepare and submit a full report of the incident to the general manager with copies to other persons as necessary. This final report will summarize all previous reports and reviews as mentioned in this section.

Index